# OCEAN ENERGY RECOVERY

## The State of the Art

Edited by Richard J. Seymour

Scripps Institution of Oceanography
University of California at San Diego
La Jolla, California

Offshore Technology Research Center
Texas A&M University
College Station, Texas

Published by the
American Society of Civil Engineers
345 East 47th Street
New York, New York 10017-2398

## ABSTRACT

This book, Ocean Energy Recovery: The State of the Art, establishes the state of the art in the full range of renewable ocean energy technologies. It includes mature technologies such as tidal energy extraction to conjectural technologies such as salinity gradients. In addition, the economics of the major systems are compared in a uniform manner, making it possible to realistically assess their economic potential. This publication provides a single source of balanced technical and economic assessments of competing technologies and should be of interest to all who are involved in the search for alternative sources of energy.

Library of Congress Cataloging-in-Publication Data

Ocean energy recovery: the state of the art / edited by
    Richard J. Seymour
    p.    cm.
    Includes bibliographical references and indexes.
    ISBN 0-87262-894-9
    1. Ocean wave power. 2. Ocean energy resources.
3. Tidal power. I. Seymour, Richard J., 1929-
TC157.O24 1992
621.31'2134-dc20                     92-32385
                                      CIP

# Table of Contents

# 1. EDITOR'S PREFACE

Richard J. Seymour[1]
Member, ASCE

The ocean, occupying most of the world surface, and the atmosphere above it intercept most of the energy from the sun -- about 80 trillion kW, or about one thousand times as much energy as is used by man globally. This energy appears in a variety of forms: as wind, driven by variations in the heating of the atmosphere; as waves, generated by wind blowing over the ocean surface; and as currents, driven principally by the wind but also caused by density differences induced by differentials in heating of and evaporation from the ocean surface.

These energy forms do not occur uniformly in any sense. Strong currents tend to form only on the western boundaries of major ocean basins. Winds over the ocean (and the resulting waves generated by them) tend to be most uniform in strength and direction in zones to the north and south of the equator. At higher latitudes, there is much greater variabilty -- ranging from long periods of calm to series of raging storms. General ocean circulation results in deep water being near freezing all over the world. Warm surface water occurs only near the equator, so that interesting thermal differences to drive power extraction machinery exist in a narrow, low-latitude band.

The rotations of the three-body system of the sun, earth and moon represent an extremely large reservoir of kinetic energy. This reservoir is tapped slowly but continuously to provide the energy to maintain the system of tides in the world ocean. These tidal flows experience hydrodynamic drag, which extracts energy from them and would eventually halt them but for the transfer of energy involved in a gradual slowing of the earth's rotation rate and the moon moving almost imperceptibly away from the earth to preserve angular momentum. The tides themselves are responses to the changing gravitational pulls of the moon and the sun as the system revolves. These gravitational forces are small, but they occur at very regular intervals so that they can excite resonant oscillations (which can be standing waves or rotary systems of progressive waves) wherever the geometry of the ocean basins will support resonance at periods close to 12 or 24 hours. The closer the local resonant frequency is to the tidal forcing frequency, the greater the response -- and the higher the tide. Tides, then, are long waves with periods close to 12 and 24 hours. We observe the tides "coming in" when the crest arrives and "going out" when the trough comes by. Tide ranges vary from near zero in the Baltic Sea to many meters along the western edge of the South Atlantic. Large tides, because of the character of the resonant waves, tend to occur close to shore and midocean tides are quite small. The most notable feature of tides, beyond their amazing predictability, is their extreme variability in range (i.e., energy density) around the world.

These two sources, solar and gravitational, are loosely interpreted as the renewable energy sources in the ocean. Neither one of them is truly infinitely

---

[1] Scripps Institution of Oceanography, UCSD, La Jolla, CA 92093-0222 and Offshore Technology Research Center, Texas A&M University, College Station, TX 77845-3400

1

renewable -- the sun will eventually run out of fusion fuel and the tidal forcing will decrease as the moon recedes. But the size of the resources compared to human needs at this time makes them attractive alternatives to burning carbon.

This book is dedicated to documenting the state of the art in the full range of renewable ocean energy technologies. As will be shown, these vary from mature (tidal energy extraction) to conjectural (salinity gradients) with every shade of development between. It provides a single source of balanced technical and economic assessments of competing technologies and should be of interest to those contemplating this resource for the first time as well as for the professional reader. All of the technology chapters have been thoroughly and anonymously reviewed by at least two internationally recognized experts. Alas, because of the small size of the ocean energy community worldwide, they must remain anonymous, without the usual listing here. Nonetheless, their contributions are greatly appreciated and had much to do with the quality of this publication.

The extraction of renewable energy from the world's oceans is a concept that has been "just around the corner" for all of my career in ocean engineering. As a graduate student -- almost exactly 20 years ago, I wrote a paper with my mentor John Isaacs (Isaacs and Seymour, 1973) that assessed the potential for ocean renewables. It is interesting to revisit this work now in light of what has happened in the ensuing two decades.

One of the first things that we did was to endorse the idea that a renewable ocean energy source, in spite of the enormous size of the world ocean, cannot be tapped without bounds. We even made some crude assessments of the size of the extractable resource by assuming that it had to be a small fraction of the natural power dissipation rates of waves, tides, currents and salinity and thermal gradients. We decided that only thermal and salinity gradients (the energy represented in all of the world's major rivers mixing with the oceans) were large enough to supply a major portion of the world power requirements. Salinity power extraction at this scale presents insurmountable environmental problems quite independent of the technical difficulties, leaving ocean thermal energy conversion (OTEC) as the only renewable resource big enough to be very significant in the world energy economy. Although the calculations were later refined (Isaacs and Schmitt, 1980), the basic findings appear to hold today.

We reviewed the status of tidal power, which was the only operational source at that time. This is a condition that persisted without much change until today. Tidal power is a growth industry, on a relative scale, because there are a few new plants on line in Russia and China. We discussed two rather revolutionary ideas, the first a floating ocean thermal power plant (OTEC) that would eliminate the problems of bridging the surf zone with a large pipe to conduct cold, deep water (Lewis et al., 1988), and the second, slant drilling an atoll for the same purpose. The OTEC plants planned today, with none yet built, are firmly fixed to shore because their primary product is electricity which has to be exported through a vulnerable cable that more than compensates for the cold water pipe advantage in floating systems. Floating OTEC systems, some very large, for non-electrical outputs are still being talked about as John Isaacs and I did 20 years ago. The atoll idea allowed us to introduce use of the nutrient enriched cold waters for aquaculture, a concept that has been since successfully demonstrated at the Natural Energy Laboratory in Hawaii (Daniel, 1985).

Our discussion of wave energy extraction included an illustration of a device for recovering the momentum transport of shoaling waves -- a concept being exploited commercially in the TAPCHAN system in Norway which remains, as of this writing, the dominant commercial wave power system operating in the whole world. It also described a wave-powered pump that, although successfully demonstrated at sea (Isaacs *et al.*, 1976), was typical of many deep water systems and would have floundered on the mooring fatigue problems had they ever been deployed for long periods. The wave power section did address the problem of energy storage for this non-constant source and described a novel scheme for producing hydrogen and oxygen by hydrolysis, storing the gasses under the ocean at ambient pressures, and extracting electrical energy through a hydrogen/oxygen fuel cell. The problems of connecting these relatively low value energy sources (because of their non-constant nature) to the electric mains is discussed in Chapter 11.

In the final paragraphs of that paper, I made the prediction that extraction of the fuels for fusion power -- lithium, deuterium or tritium -- would eventually be the the greatest energy source from the ocean. Although not strictly a renewable energy source, this could, of course, eventually happen. Because we seem to have come no closer to practical fusion power than we were 20 years ago, it appears that I will probably not be around to be proven wrong. The fusion analogy is probably very appropriate to ocean energy. In the one case (fusion) we have a very difficult -- perhaps impossible -- series of technical breakthroughs to accomplish, with no real guarantees that we can afford the power if it is generated. Yet, internationally, we have funded fusion research for 40 years at levels orders of magnitude higher than the highest ever achieved for ocean energy. In the other case (ocean renewables) we are within a few years of having operational plants on line in small, local technologies (tides and waves) and potentially large, global sources (OTEC) and yet the funding base for research and development has all but disappeared.

The strong interest worldwide in the ozone depletion and greenhouse gasses generation problems has somehow as yet failed to reignite interest in the renewable energy resources of the ocean. It is hoped that this book will be a contribution towards establishing the feasibility of some of these more mature technologies and lead to a resurgence in the development of environmentally benign alternatives.

**References**

Daniel, T. H., "Aquaculture Using Cold Water," *Proceedings*, Marine Technology Society, OCEANS '85, San Diego, CA., Vol. 2, 1985, pp. 1284-1289.

Isaacs, J. D., Castel, D., and Wick, G. L., "Utilization of the Energy in Ocean Waves," *Ocean Engineering*, Vol. 3, 1976, pp. 175-187.

Isaacs, J. D. and Schmitt, W. R., Ocean Energy: Forms and Prospects," *Science*, Vol. 207, No. 4428, 1980, pp. 265-273.

Isaacs, J. D. and Seymour, R. J, "The ocean as a power resource. *Int. J. Environmental Studies*, Vol. 4, 1973, pp. 201-205.

Lewis, L. F., Van Ryzin, J. and Vega, L., "Steep Slope Seawater Supply Pipeline," *Proceedings 21st International Conference on Coastal Engineering*, ASCE, Malaga, Spain, June 20-25, 1988, pp. 2641-2654.

# 2: THE STATE OF THE ART IN TIDAL POWER RECOVERY

## J. Gavin Warnock[1]
## Robert H. Clark[2]

### Abstract

*The State of the Art in tidal power recovery is, in 1990, well advanced and no significant technical or engineering constraints apply. A number of notable tidal power developments have been accomplished and energy has been extracted from simple tidal mills throughout the ages. Any constraints that might arise from economic, financial and environmental considerations today are those associated with development of almost all electrical energy sources.*

*Basic concepts, tidal power technology and the practical achievements in this field will be described. The state of the art employed in the engineering design of major proposed tidal power developments is taken as the basis of this chapter. Intensive design consideration has been given to the application of advanced achievements in marine offshore construction and in the manufacture of large hydroelectric generating units and power plants. Experience in these fields allows for convenient transfer of technology to tidal power recovery.*

*Construction method and means of shortening schedules are critical to minimizing cost. The best endeavours on both counts do not yet achieve an economic outcome which meets the criteria for normal financing of power generating facilities. Secondary benefits from ancillary uses are an important factor in planning for tidal power but rarely contribute significantly to reducing energy cost. Future development of tidal power will be dependent on a return to low discount rates combined with a growing recognition of the human value of security from diversity of energy sources benign in their global effects.*

*Environmental and ecological effects are all important factors in considering the acceptability of energy developments. A tidal development, by its nature, affects the relatively delicate inter-tidal zones of the inlets which create the near resonant conditions and the basin storage necessary for its operation. The scale of changes arising from any sizable tidal power development will almost certainly be substantial and will need to be assessed on a site specific basis.*

## Introduction

Of all renewable energy resources available to mankind, tidal power recovery harnesses the most predictable and irresistible forces of nature; gravity and centrifugal action. The origin of the forces is "extraterrestrial" and unaffected by the earth's atmosphere, depending principally on the relative positions and movements of planetary bodies.

[1]*Chairman, LE Energy Limited, 9 Beechwood Mains, Edinburgh, EH12 6XN, UK.*
[2]*Consulting Engineer, 8567 Sunsum Park Drive, Sidney, BC V8L4T6, Canada.*

4

The influence these forces have on the oceans of the globe, combined with the fortunate shaping of coastal geographic features which amplify tidal effect, have provided a substantial number of sites around the world where tidal energy can be recovered. This depends largely on concentrations of flow in and out of confined tidal basins and the range of ocean levels which occur there with unerring predictability and regularity.

This chapter sets out the basic concepts of tidal power recovery with particular recognition of the variability of ocean effects, season by season and day by day, out of phase with the solar day/night cycle to which human activity is tied Although predictable, tidal energy output is cyclical and therefore of diminished value when it does not occur in phase with demand.

Ways and means of enhancing the availability of tidal power have presented challenges, the solutions to which tend to conflict with the fundamental objectives of minimising the investment cost of the facilities necessary to harness the energy source. Growing electrical demand in regions accessible to tidal power generation output enhances the ability of supply systems to absorb cyclical output and, over time, has improved the status of this energy source as a competitive alternative.

The technology of hydroelectric turbines and generators, too, has advanced in a manner favourable to tidal applications. Very large machines can be constructed which, in spite of relatively low level differences across a tidal barrage, can develop significant outputs. Adaptation of conventional design approaches has brought a fine degree of specialisation in tidal applications.

The generating plant and associated switching and transmission equipment form typically one third to one half of the tidal power development and its cost, with civil works representing the major proportion. Within recent time, progress has been made to closer integration of plant and the civil engineering structures containing it.    Employing methods now commonly accepted in the off-shore marine construction field, floated modules, prefabricated in calmer in-shore facilities, are used to form tidal barrages without dewatering the site. Substantial reductions in schedule and cost savings are achieved.

Technology has met most of its challenges but economic and environmental constraints, which are to some degree interdependent, remain. The state of the art in tidal power recovery has advanced to provide improved knowledge of the impacts of these constraints.   By and large adverse environmental outcomes can be mitigated although tidal basin contours will be permanently changed by sediment movements under the changed regime conditions.   The remaining powerful challenge is to match the overall benefits of tidal power recovery, as an indigenous truly renewable energy resource, to the economic realities of those jurisdictions enjoying the privilege of high tidal ranges.

## Basic Concepts of Tidal Power Recovery

Theoretically, any tide can produce power in some quantity from the energy otherwise dissipated during its ebb stage. For any tidal power scheme, it is the range of the tide and variations in that range which determines the potential head available for exploitation at a particular site. The amount of rise and fall of the tide is not the same everywhere; it varies from almost nothing to a range of over 17m in some localities. Since the laws governing tidal fluctuations are basic to the issue of tidal energy exploitation and utilization, the basic concepts and terminology are described in this section.

### Tidal Origins and Effects

The ocean tide is produced primarily by fluctuations in the resultant of the forces of gravity and of centrifugal action caused by Earth's rotation and the continuous and repetitive changes in the relative positions of Earth, the moon and the sun, as depicted by Fig 2-1 [Dean, 1966]. Because of their relative size and distance, the tidal effects caused by other celestial bodies are minute. Even though it is considerably smaller, the moon's proximity to Earth allows it to exert a much greater influence on the tide than does the sun, since attractive force is proportional to the inverse square of the distance between the bodies; in theory, the magnitude of the solar tide is about 46 per cent of the lunar.

The tide in the various oceans progresses through them as an undulation. For example, the tide runs up the Atlantic Ocean as an undulation from its south end between South America and Africa to its north end between Canada and Europe [Dawson, 1920].

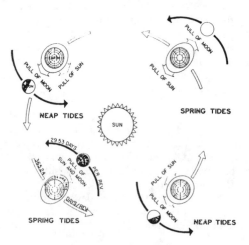

Fig 2-1  Origin of tidal effects on the earth

The range of these oceanic tides is usually not more than about one metre. On the shore of an ocean, the water level rises and falls, a fluctuation which occurs twice each day or, more precisely, twice every 24.84 hours, the apparent period of the rotation of the moon around Earth. Primarily due to the gravitational pull of the moon, the water level gradually rises until it reaches its highest point, termed "high water", and then it falls to a low level, or "low water" over a period of 12.42 hours, after which the process is repeated. The difference between the consecutive high and low waters of this semi-diurnal tide is called the "range" of the tide. Because of the excess of 0.84 hours over the solar day, the tidal rhythm continuously moves into and out of phase with society's activities which are geared to the solar cycle.

Because the paths of the moon around Earth and of the Earth-moon system around the sun are elliptical, the distances between these bodies, and therefore the attractive forces acting upon the oceans, vary continuously. Nevertheless, they can be computed precisely. Obviously, because of the relative positions of the planes of rotation of the moon and Earth and their elliptical paths, the relative influence of the sun and moon are different on different regions of Earth. Thus, at every locality in every ocean, there will be found a well-defined variation in the range of the tide which usually recurs twice in the course of each month but this variation may be different from region to region; for example, there may be a large inequality in either the high water levels or the low water level, or both, and occasionally the tide may become diurnal. Such tides are prevalent, for example, along the Pacific coast of North America and in the Caribbean. There are some locations, such as on the Vietnamese coast, where the tide is **diurnal** in character, i.e. with only one high water and one low water during the day. However, whatever the character of the variation may be, it occurs in an orderly fashion and is predictable.

Principal Tidal Types

Taking into account the variations in their originating forces, the tides may be grouped into the following principal types: synodic, anomalistic and declinational tides. Although one variation may be dominant, the others are never entirely absent, and may be of sufficient magnitude to yield a "mixed" tide.

The synodic tide - occurs when the dominant variation in the range of the tide takes place twice a month, the range being greater at new and full moon and less at the moon's quarters. Referring to Fig 2-1, when earth, the moon and the sun are in line, which occurs about twice a month, the attractive forces of both the moon and the sun are in conjunction and produce the high or "spring" tide. About eight days later, the sun and moon are at right angles with earth and partly neutralize each other's gravitational influence on the oceans so as to produce a tide with a low range or "neap" tide. When the moon is new or full and closest to earth, i.e. during the equinoxes in March and September, the tidal range is particularly large. At some locations it may be as much as 12 times the neap tidal range. Such extreme differences would have an important bearing on the design and operation of a tidal power plant. The length of the synodic month varies because of irregularities in the moon's movement as a result of the sun's attraction, but its average value is 29.53 days.

An anomalistic tide - results when the variation in range accords with the moon's distance and takes place once a month. The path of the moon as it revolves around the earth is elliptical but earth is slightly off centre of that elliptical path. Therefore, during each revolution of the moon around earth, there is only one point where the moon is nearest earth and that point is known as the moon's "perigee". The period of time from perigee to perigee is called the anomalistic month which has an average length of 27.55 days. Although there are no tides which are purely of the anomalistic type as there are of the synodic type, there are regions, such as the Bay of Fundy on Canada's eastern seaboard, where this variation in tidal range with the moon's distance is distinctly greater than the variation from springs to neaps [Dawson, 1920].

The declination tide - results when changes due to the moon's declination (which makes the two tides of the day unequal in range) is so large and obvious that all other features of the tide are obscured. The period of time that elapses between the successive times that the sun crosses the equator in the same direction is, of course, 27.32 days, a period known as the "tropical" month. In regions where tides are dominated by the moon's declination, the variations which result are greater than from any other movement of the moon considered separately. When the sun and moon are on the equator, the two tides of the day will be equal in range but, when they are not at the appropriate declinations to balance their respective declinational influences, there is a marked "diurnal inequality".

Variation in Tidal Effects

The cyclicity of the tides is evidenced in amplitude as well as frequency. The spring to neap cycle leads to a 14 day cycle in which the tides vary in amplitude by as much as +/-400%. Furthermore, because the moon's orbit is an ellipse, successive spring to neap cycles can vary in amplitude by typically +/-15%.

The inclination of the earth's orbit to the moon causes a semi-annual effect with higher spring tides in March and September, of the order of +/-11%. Further complex gravitational interactions lead to a 19 year cycle with a +/-4% effect on tidal amplitude.

Variations in tidal amplitude have an even greater effect on the potential for energy extraction, which is proportional to the square of the amplitude.

Above all it must be realised that the major difference between tidal energy and other renewable sources is that the variation in the amplitude and frequency of the tides and their stage at any given time are predictable with a considerable degree of accuracy.

## In Shore Tidal Effects

The tides produced by celestial and gravitational forces have almost always a symmetrical undulation at points distant from the earth's land masses. The tidal range in the open ocean is about 1 metre. This undulation causes, at its summit, high water and, at its trough, low water and these can be affected substantially by the configuration of the shorelines where the tidal effects impinge.

On entering an estuary, bay or inlet the form of the undulation is subjected to effects more closely related to laws of hydraulics and wave motion than to celestial effect [Ippen and Harlemann, 1966]. The sea level is raised and, depending on the configuration of the estuary or inlet. The effects of shelving, funnelling and reflection can amplify the tidal range to as much as 13 metres at some sites.

The period of the semi-diurnal tide makes the wave-length much longer than the physical length of estuaries and inlets occurring in natural geography. Lengths of about one quarter or half of the tidal wave length do impart, however, a resonance response which can increase the amplitude of the tide by factors of four or more. Fig 2-2 shows the estuary lengths of the Bay of Fundy inlets which create an amplification factor of about 3.5.

Sites, where favourable conditions arise for recovery of tidal power from the oceans, are to be found, then, throughout the world where shoreline configuration produces substantial amplification of the continuous and periodic undulation of the ocean's surface (see Fig 2-5). Optimal sites are mainly determined by the degree of amplification effect but also by the convenience offered by the shoreline configuration for the location and construction of the facilities for tidal power recovery.

## Selection of Optimal Sites

The simplest solution to the recovery of tidal energy and its conversion to electricity is that used in the construction and operation of the ancient tidal mills. A dam or barrage is constructed across an estuary, equipped with turbo-generators and with sluiceways so as to allow the basin to be filled on the rising, or flood, tide. At high tide, the sluices are closed giving rise to a head with respect to the ebbing water seaward of the barrage. When the head differential has increased to about half the value of the tidal range, i.e. with the seaward water level at about mean sea level, generation is started. Operation then proceeds with the flow from the basin to ocean until the head on the turbines has reached the minimum value under which the turbines are capable of developing power. At this point, the turbine gates are closed and, after a short waiting period, the ocean level rises to that in the basin. The sluice gates are then opened and the basin allowed to fill with the rising tide. This mode of operation of the simple, single-basin scheme is known as "ebb generation".

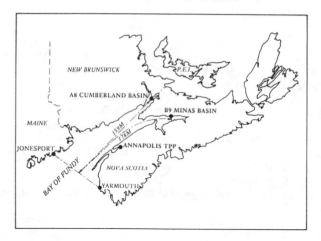

Fig 2-2  Tidal resonance (Bay of Fundy)

The single-effect (one-way generation) turbines could be arranged to discharge into the basin during the rising tide.  However, operation in the ebb-generation mode will yield more energy than generation on the flood tide since the former operation takes full advantage of the greater volume in the upper levels of the basin at higher heads.

Modern turbine technology has made it possible to use the same turbo-generator set to produce electrical energy on both the rising and the falling tides, viz. "double-effect" turbo-generators.  The La Rance and Jiangxia tidal power plants are capable of this double-effect operation.  The generating units are capable of generating, as well as pumping in both directions and also to operate as orifices in both directions to assist the sluices, viz. six modes of operation are possible.  The attraction of such two-way generation is that it provides not only for the production of electricity during a greater portion of the solar day than single-effect, but it also permits greater operational flexibility.  Fig 2-3 illustrates the layout of a single-basin, single-effect, ebb-generation scheme and a single-basin, double-effect scheme and the general characteristics of their outputs.

It will be noted that, with double-effect operation, the sluices must be opened sooner near the end of both the ebb and flood generation periods in order to reduce the basin level quickly to provide for the flood generation or to fill the basin as quickly as possible to the start of ebb generation.  Thus, for any given number of turbines and sluices, slightly less energy would normally be produced by two-way or double-effect generation than by single-effect (ebb generation) only.  Under the former mode of operation, the operating heads are generally lower than for single-effect operation.  On the other hand, the increased heads developed under higher tidal ranges could be more fully utilised with double-effect operation so that, as the tidal range increases, the latter operation could generate more energy per tide with the same number and size of sluices and turbines.  However, with two-way

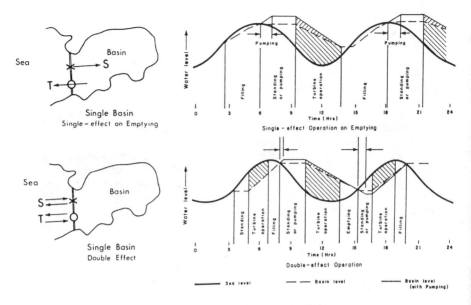

Fig 2-3  Energy output characteristics for single basins

generation, the turbines would operate somewhat less efficiently at any given head than for single-effect or uni-directional flow since optimisation for the latter condition must be compromised to obtain an optimum design for double-effect or two-way flow.

Tidal power schemes generally benefit from pumping with the generators, operating as motors, driving the turbines as pumps at times when the tidal levels is high and little different from that in the tidal basin.  The pumping is at low head and the operation increases the effective volume of water for generation at high head. Typically, an increase of 5 to 15% above that for ebb generation can be achieved. Pumping also enables some limited retiming of output.

In summary, a single-basin, double-effect development is capable of many modes of operation between that for production of energy only and the mode for the production of dependable peak power.  The development can be programmed for an operation that will maximize the revenue from the production of its energy throughout a solar day.

A coastal configuration favourable to the creation of two basins that could be hydraulically linked would enable the production of continuous power from the tides as shown in Fig 2-4. [International Passamaquoddy Engineering Board, 1959].  The water level in one basin is held high and that in the other low, with generation always in the one sense, from high to the low basin.  The former would be filled during high tide through one set of gates and the latter emptied during low tide through another set of gates. Although the output would be continuous, its energy production would be lower than that from a single-basin scheme, using either of the basins. Moreover, because of the additional civil works required, it would tend to be substantially higher in capital cost than a single-basin scheme, and the cost of electricity would therefore be correspondingly higher.

A paired-basin scheme, i.e. two single-basin developments that are electrically interconnected would also afford some flexibility in operation to most market demands. Advantage could be taken of any differences in time of high water between the sites. With the objective of obtaining a more uniform combined peak power production, one plant would be operated with a high basin level and the other with a low basin level. A number of other variations in operation of single, double and paired basins to achieve specific goals have been considered [Bernshtein, 1965].

It will be evident from the foregoing brief descriptions that, in addition to a sufficiently high tidal range, a site should also include a natural bay with an adequate area and volume and be so situated that the operation of the plant will not significantly reduce the tidal range. There are relatively few locations in the world where esturial configurations promote sufficiently large increases in tidal range and, at the same time, enable a sufficiently large area of the estuary to be isolated by a relatively short and shallow barrage.

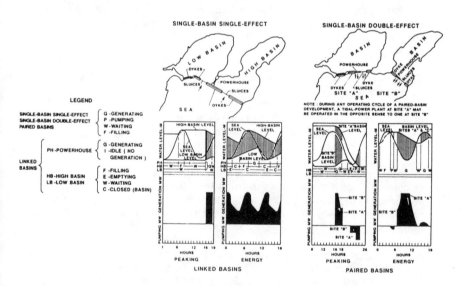

Fig 2-4  Multi-basin schemes and characteristic outputs

Fig 2-5 shows the approximate location of sites on which some activity has been reported during the past decade or so. Since the available power varies as the square of the range and the area of the basin that can be created, these are two of the major factors that determine a suitable site. However, there is a wide array of other considerations that come into play as a result of oceanographical, geographical, technological and economic factors. These factors are seldom simultaneously favourable.

Although hydroelectric and tidal-electric power developments have many characteristics in common, there are some basic difference in approaches to their design and construction. For example, the most economical design for the former type is based on the hydrological characteristics of the river, such as minimum recorded flow, flood flows, storage possibilities and head. For a tidal development, however, the design can be varied to produce the desired capacity.

Thus, the output of a tidal plant is determined by the following interrelated factors:-

i. useable head which varies continuously with the tidal regime and is modified by the fluctuation of basin levels resulting from the operation of the plant;
ii. the area of the tidal basin;
iii. the capacity of the sluices used to fill or empty the basin;
iv. the capacity of the generating units; and,
v. the method of operation selected.

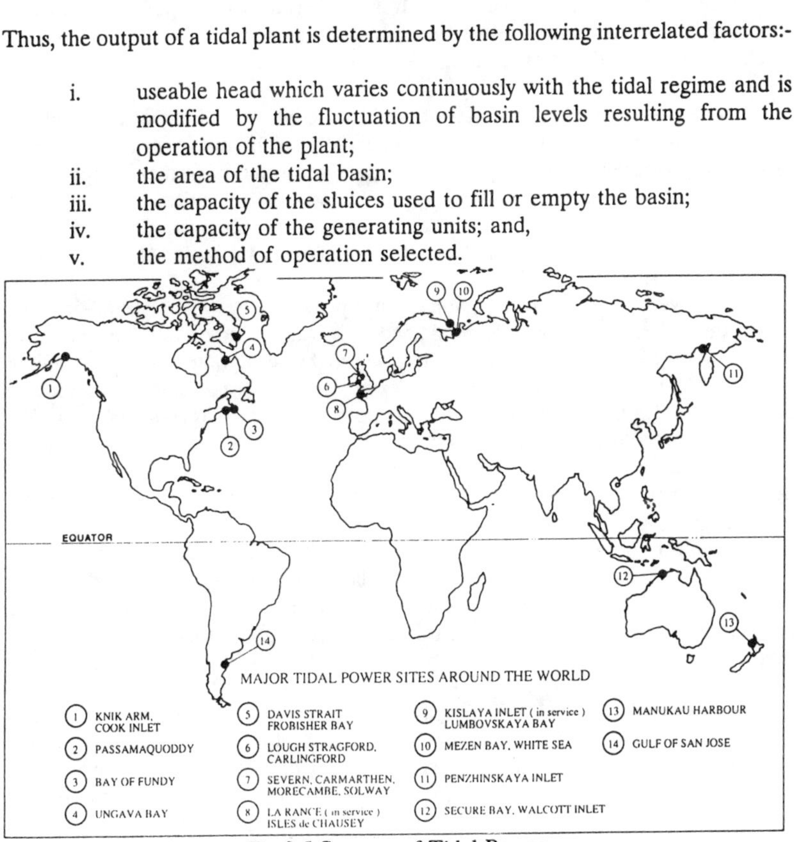

MAJOR TIDAL POWER SITES AROUND THE WORLD

| ① KNIK ARM, COOK INLET | ⑤ DAVIS STRAIT FROBISHER BAY | ⑨ KISLAYA INLET ( in service ) LUMBOVSKAYA BAY | ⑬ MANUKAU HARBOUR |
| ② PASSAMAQUODDY | ⑥ LOUGH STRAGFORD, CARLINGFORD | ⑩ MEZEN BAY, WHITE SEA | ⑭ GULF OF SAN JOSE |
| ③ BAY OF FUNDY | ⑦ SEVERN, CARMARTHEN, MORECAMBE, SOLWAY | ⑪ PENZHINSKAYA INLET | |
| ④ UNGAVA BAY | ⑧ LA RANCE ( in service ) ISLES de CHAUSEY | ⑫ SECURE BAY, WALCOTT INLET | |

Fig 2.5 Sources of Tidal Power

Because of the abundance of water available to a tidal plant, design emphasis is more on maximum output than on maximum efficiency, the latter being more significant for hydroelectric developments. Also, because of the continually changing head, it is impossible to determine a priori the dimensions of the equipment and the magnitude of the installed capacity for a tidal electric development. Optimisation is, therefore, a fundamental requirement in the design of a tidal power development.

## Power and Energy Generation from the Tides

The selection of an optimal tidal power site provides the definition of the ranges of potential operating head which will occur from tidal effect and of the volume of the land side basin(s) from which ebb flow can be extracted. It has been noted that energy can be extracted from tidal power potential on either the flood (inwards to the basin) or ebb tide (from the basin). It is, furthermore, quite possible to extract energy from both directions of flow. Economic considerations influence the selection of ebb or two directional operation.

A further variation in plant design is offered by the inherent ability of the low head turbine to function in a pumping mode when "driven" by the generator operating as a motor powered from the system. This does add some degree of complexity to the tidal plant but recent conclusions support the economic justification for pumping at least during the slack water period at the end of the ebb cycle. [Severn Barrage Project; 1989]. Net energy output may be increased in the order of ten per cent.

Low head hydroelectric power equipment has proven to be well adaptable to tidal power duty where maximum potential heads might be expected to fall in the range of, say 5 to 13 metres. This range is covered by axial flow turbine design of relatively high specific speed. Precedent experience has established hydraulic and cavitation performance as well as the economy of large diameter relatively low rotational speed units. Variable turbine blade angles allow a wide range of heads and flows to be accommodated at reasonably high efficiency. The design and development of low head hydroelectric equipment is well advanced and mature.

It is, however, notable that tidal power technology has been more drastically influenced by advances in the specific application of hydroelectric turbine and generator design than by any other element of this mode of power extraction. The past 25 years has seen a notable consolidation in the state of the art as applies to tidal power.

The emergence in the 1960's of the horizontal shaft axial flow water turbine and direct driven generator housed in an underwater "bulb", as shown on Fig. 2.6(a), brought a new level of technical perfection to the overall form of tidal power plant. The ability to convey flow through the barrage without change of direction brought welcome simplification to both the equipment and to the structure in which it is installed. It also led to more ready adoption of modular caisson construction (dealt with later in this chapter), this being the other highly significant advance of the past 25 years.

While bulb type units have been in use since the mid 1930's, the wide application, mainly to run-of-the-river hydroelectric plants in Europe and only more recently in USA and Japan, has taken place concurrently with a renewed interest in tidal power. The first application to tidal generation at La Rance, France no doubt brought apparent economies which gave impetus to the justification for that project. [Gibrat, 1966].

Understandable reluctance to locating the generator in a capsule or nacelle within the water stream, thus imposing constraints on physical design and on operational convenience, brought forward alternative ideas for achieving the basic objective with more conventional treatment of generator design. Open pit arrangements with flow carried around either side of a central pier and inclined tubular units with the generator drive shaft projecting from the inlet water passage have had some low head river applications but none yet to tidal power generation plants. [Seoni, 1976]. For smaller tidal power schemes with modest unit sizes, a geared bulb machine in a pit configuration could provide an economically attractive.

One alternative form of tidal power turbine in the general category of axial flow horizontal shaft machines is the Straflo (Fig 2.6 (b)), already successfully applied to a commercial tidal project at Annapolis, on Canada's Bay of Fundy. The concept of a hydraulic turbine where the flow passes from intake to discharge with the minimum of diversion -- even less than that caused by the bulb surrounding a submerged generator -- had challenged designers for many years. The solution

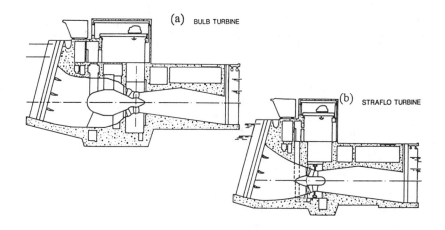

Fig 2-6 Horizontal shaft axial flow turbines

obviously requires the turbine flow to pass concentrically through the rotating element of the generator [Holler & Miller 1977].

The key to achieving such a design was the development of sealing devices which could operate on relatively large diameters and at high surface velocities. Reliability had to be an essential requirement as the peripheral water seals were of necessity in the region of the generator windings. Moisture proofing of coils and connections (or possibly even encapsulation), made possible with modern technology, contributes to the practical feasibility of generator design. The turbine rim seals of the Straflo type must maintain a gentle outward flushing action for cooling and cleaning; a positive outward leakage of generator casing air minimises take up of moisture in the interface area.

Hydraulically the state of the art Straflo turbine has reached a standard of near perfection with only stationary struts and an essential central nacelle offering any obstruction to the flow in the water passages of the turbine. In the evolution of the design, concepts were studied, designed and tested at model scale which would have allowed peripheral bearing support of the generator rotating mass at the outer rim rather than on the conventional central shaft. Such truly "straightflow turbine" design has not yet reached maturity [Braikevitch:1979].

It will be apparent that support and driving of the generator at the runner rim periphery on the turbine presents challenging design problems which are most easily accommodated with a robust and rigid connection. Relatively small Straflo turbines have, however, been built with rim support on the outer tip bearings of variable pitch turbine blades. Such double regulation has provided for improved efficiency performance at Hoengg, Switzerland (and Weinzodl, Austria), but large-scale tidal applications will most likely be fixed turbine blade, adjustable wicket gate, single regulated designs. The overall hydraulic performance is comparable to that from the alternative bulb-turbine designs available for tidal application with variable runner blade pitch and fixed guide vanes [Severn Barrage Project: 1989].

It has been noted that the ability for tidal power plant to pump under near balanced water levels can be important. For reasonable efficiency performance double regulation with variable runner blades and adjustable guide vanes is required. No large Straflo turbines have yet been built and tested with this configuration. Where pumping is found desirable current practice favours selection of double regulated bulb machines.

Regulation of Tidal Turbines

The regulation of tidal power turbines for prime performance has provided a field for some adaptation of conventional approaches. Relatively wide ranges of net operating head and flow have to be accommodated over the tidal cycle, and turbine runners with variable blade pitch are usually employed to give good efficiency throughout the operating zone. Double regulation with both adjustable wicket gates and variable pitch runner blades, following Kaplan design principles, gives marginal improvement in overall efficiency. However, in those cases where pumping does not predominate, the state of the art in design of tidal turbines adopts single regulation, with adjustable runner blades only, in a "Kapeller" configuration (Fig 2.7). The elimination of operating gear for wicket gate adjustment allows a compact machine more readily adaptable to caisson installation. It will be seen that tidal barrage economics are strongly influenced by cost savings resulting from modular floated-in construction that this configuration permits [Severn Barrage Project 1989].

Turbine design has a further strong influence on cost arising from the requirement for submergence below minimum water levels on the ocean side of the barrage. Increases in the depth of setting can lead to substantial excavation cost penalties. Deeper settings do, however, control cavitation effects which inhibit performance and can cause damage to plant. Cavitation limitations combined with the desire to optimise energy output per tidal cycle lead to turbine centre line settings and overall excavation requirements calling for careful balancing of construction cost/energy output parameters. A high standard of cavitation performance is a prerequisite for economic tidal turbine applications from both initial capital expenditure and long term operating cost considerations. State of the art standards are high in this regard and have been assisted, in recent years by advances in the capability for three dimensional asymmetrical modelling analysis of rotating flow channels.

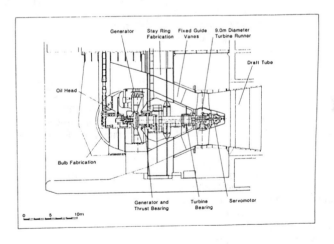

Fig 2-7 KAPPELER Turbine Generator

## Large Generating Units Bring Economies

Perfection of plant design has had to accommodate an economic necessity for machines of large physical size and generation output capability. The early tidal power plants were equipped with units of sizes well below the optimum for energy output cost and trends to larger and larger machine diameters have matched or exceeded that in the hydroelectric industry at large. This is so because major tidal power installations can involve numbers of generating units an order of magnitude more greater than those required for conventional developments. This can lead to acceptance of unusual commitment to large scale, purpose built manufacturing and transportation facilities removing at least some of the usual limitations.

Run-of-river hydroelectric plants have been equipped with bulb units of 8.2 metre diameter which have been almost matched by the Annapolis Straflo tidal machine at 7.6 metre. The state of the art represented by the careful considerations of the Severn Barrage studies would suggest 9.0 metre runner diameter is a current "limit" influenced by site and turbine setting considerations. It should be noted, however, that this is an increase from 8.2 metre adopted only 3 years before and the 9.0 metre size does not represent an absolute limit. The 9.0 metre machines would operate at 50 rpm, and under the tidal conditions at the Severn, would develop a rated generator output of 40MW.

## Implications for Generator Design

It will be apparent that the horizontal shaft axial flow hydroelectric unit imposes particular constraints to the design of the electrical generator component. The concentricity of the water flow with the axis of rotation of the machine calls for it to flow either around the generator casing (the bulb type) or through the centre of the rotor (the Straflo approach).

Other configurations have been noted including the tubular inclined shaft machine with the generator, located away from the water passages, driven by the shaft passing through the water passage casing as it bends to the horizontal inlet transition section. Units of this type have been given comparative consideration with other forms of tidal power plant of "limiting" dimensions, [Severn Barrage Studies 1986], but are no longer seriously considered.

The constraints of locating, operating and cooling generators within steel enclosures in the water stream do appear to have become less onerous as design ratings and physical dimensions have increased. Some large units have been built with gear trains to allow step-up of turbine speed by ratios of the order of 10:1 [Hadley 1990]. The higher generator speed has, with reduction in diameter, increase in core length and in number of pole pairs, made it easier to accommodate the machinery within the bulb. Proven designs of the double epicycle gear boxes required are limited to about 25MW generator drives. Interest in gear driven tidal power units is then limited to smaller power outputs. This can lead to lower costs and improved hydraulic water passages.    However, it is likely that bulb-type, direct-driven generators will meet tidal power needs where unit capacities are at the upper limits.

One potential disadvantage of the constrained diameter bulb type generator is its limited inertia (H factor) although this is not necessarily of operational concern. The alternative rim type unit driven by the Straflo turbine does not suffer this limitation.    It has a diameter and flywheel effect of the same order as a conventional hydroelectric generator. Electrical performance may be expected to be marginally better. As noted, however, there is some disadvantage since it must function in moist areas which leads to a need for special operating routines.

State of the art electrical design permits assynchronous output from generators to be regulated to match the system frequency needs through conversion facilities. This has provided another variable for plant design consideration. Rotational speed varying from synchronous by some +/-5% can give appreciable improvement, say about 2.5%, in overall tidal cycle energy output. The cost effectiveness of this provision is, however, questionable [Severn Report 1989].

## Economic Pressures on Design

The form of the modern tidal power generating unit and the measures employed to minimise the cost of construction of tidal barrages results in the mechanical electrical plant components being a relatively high proportion of the overall project cost.    Operation and maintenance of the power components also form an appreciable proportion of this cost account. As economic challenges to tidal power development from other sources of electricity generation continue to be severe the state of the art design has been "driven" by cost considerations.  With the physical size of generating units bringing economies both directly and indirectly, through the convenient construction integration with floated-in caissons, costs of tidal power recovery facilities are being effectively minimised.

## Tidal Barrage Retention Facilities

A tidal barrage comprises three primary elements: a reinforced concrete power plant, similar (but simpler) sluiceway sections and embankments constructed of rock and earthfill, with or without impervious cores, to form the controlled tidal basin. Structures for navigation locks or other needs may also be required. The configuration and contours of the barrage site strongly influence the proportions of these elements. An economical tidal site will be likely to have:-

-       good foundation conditions limiting the need for elaborate preparation before placing the structures;

-       a depth of water at low tide to ensure proper turbine operation; and,

a relatively short length of embankment, ie. the "inactive" and comparatively low cost section of the barrage.

When it is realised that some of the proposed tidal power developments will have barrage lengths of up to 20 km, and even longer, the selection of a construction technique will have a significant impact, not only on the direct cost of the civil works but also on the construction schedule.

The structures of a hydroelectric development are normally constructed within cofferdams so that the powerhouse and sluiceways of an earth or rock-fill dam can be erected "in the dry". This technique was used at La Rance tidal power development between 1961 and 1966 (see Fig 2-8). However, temporary cofferdams accounted for about 30 per cent of the project's civil works costs [Cotillon, 1974], and led to closing off the estuary for two or three years, which would not be considered environmentally acceptable today. During the past 30 years there have been significant advances in the design, construction and placement of concrete caissons for marine installations. This type of construction has found application in many types of marine structures. Floated-in caisson sluiceway units were successfully placed in the 1960's in the construction of the storm surge barriers in the Netherlands, as part of the Deltaworks Plan.

In the late 1960's, the Russian engineers installed an experimental tidal power plant at the entrance to Kislaya Guba. A simple two-unit turbine and sluiceway caisson was fabricated near Murmansk and towed about 100 km to the site. The caisson was placed within a few millimetres of the correct vertical position and only a few centimetres from the proposed horizontal position [Bernshtein; 1974]. Floated-in construction technology was very considerably advanced during the 1970's and 1980's in connection with the offshore oil exploration and production fields in the North Sea.

Fig 2-8 Aerial view of construction cofferdams and the dewatered site behind cofferdams for the powerhouse and dyke of the La Rance tidal power plant

There will always be site configurations where *in situ* construction may be more economical, but the state of the art clearly favours caissons built in shore-based facilities, floated to the barrage location and ballasted down onto prepared foundations.

The major advantages of this method of constructing a tidal barrage are:-

- minimum environmental impact, particularly on sediment movements during construction;

- significant construction work can be carried out remote from the barrage site, avoiding the very large concentration of resources required for *in situ* construction; and,

- the programme of installation can be accelerated, yielding economies through a substantial reduction in the schedule of construction.

As already noted, the large diameter, horizontal axial-flow units now available suit floated-in caisson design. In addition, the sluiceways, which also function as horizontal waterways through the barrage, can be built and placed in the same manner. This method of construction is even finding a place in the building of large run-of-the-river hydroelectric plants such as Vidalia on the Ohio River (USA). (see Fig. 2-9 and Fig. 2-10).

Fig 2-9
Vidalia Power Plant in place on prepared foundations.

Fig 2-10
Floated-in prefabricated steel hydro-electric power station under tow.

Within the state-of-the-art, caisson units of 80 to 90 metres in plan with a draft of the order of 21 metres (85,000 tonnes displacement) are being considered for estuarine construction and placement. While marine exposures would be generally similar to those at open sea sites, less aggressive wave and wind conditions are expected at tidal barrage locations. Flow velocities at barrage closure would be similar to those experienced at coastal sea dykes in large estuaries for which successful closure procedures have been developed.

The large-scale tidal power developments under study, with barrage lengths ranging up to 20 kms, may contain 100 to 200 or more turbo-generator units with, perhaps, four units per caisson and about half as many sluiceway as power caissons. The principle of dock-built, modular component construction can be extended to the simple closure portions of the barrage which will often match in length the sluiceway sections, depending entirely on the site conditions and configuration. It would not be unusual, however, to have as many as a hundred or so floated-in modules placed in their final setting determined by the contours at the barrage site. Overwater marine operations can prepare, by dredging, filling, vibration, compaction and grading, the foundation pads on which the caissons are positioned, secured and sealed in place.

A major incentive to construction with floated-in components is the ability to carry out structural civil engineering works in protected inshore dock conditions. The placement of equipment and fitting out may be undertaken at the shore facility, or large assemblies of turbo-generator equipment may be barged to the caisson that has already been set down on the barrage line, and placed by the very large floating, lifting derricks now available. With adequate building facilities, production rates can be achieved to allow an almost continuous tow out and placement process, permitting much shorter overall construction schedules than would be possible with *in situ* building in the dry, behind cofferdams. As the overall cost of a capital intensive project is greatly influenced by its construction time, the speed of completion and closure of the retention facilities is undoubtedly one of the major keys to economic viability.

Fig 2-11 Sketch showing installation of preassembled turbine generator units.

### Existing Tidal Power Developments

### La Rance, Brittany, France

Prior to La Rance development, tidal plants had been planned on the basis of vertical shaft hydroelectric turbines and generators. It has already been noted that the progression to horizontal shaft, axial-flow, bulb-type turbo-generators, of which La Rance was the prototype example, revolutionised large-scale, tidal power development. Twenty-four 10MW units operate with 8.5 metre tides.

The barrage site required only a very limited "inactive" dyke section which was constructed with the powerhouse "in the dry" behind sheet-piled cribs and concrete caissons Fig 2-12). The sluiceway section was also constructed "in the dry" prior to the placing of the cofferdams for the powerhouse. By today's standards, the construction method led to relatively high costs, as did the selection of relatively small generating units which were also designed for double-effect operation and pumping in either the ebb or flood direction.

A comprehensive, impressed-current cathodic protection system was incorporated with the scheme and has proven very successful in preventing corrosion of the stainless steel runner blades and the other submerged metal parts. With the exception of a design weakness in the generator stators requiring minor modifications, the turbo-generators have proven reliable and maintenance requirements modest. It has been concluded [Cotillon, 1979] that La Rance is not only a source of valuable lessons, but also "an economically attractive power station, since the standard cost of tidal energy generated in 1976, for example, was equal to the average cost of a nuclear kWh and also equal to the average thermal kWh cost". Much depends on the discount rate used for capital employed for this to be true.

Fig 2-12  France undertook the first large tidal development at La Rance in 1986.

## Kislaya Guba, USSR

At a unique site on the coast of the Barents Sea requiring a sea cut-off of only 50 metres, a concrete caisson containing a power plant and sluiceway was floated into place from a dock area near Murmansk, about 100 km distance. The particular location of this scheme was selected primarily to require minimum work to close off the bay which would become the tidal basin for this experimental station. Although built for two 400 kW bulb units, the plant has only been equipped with one unit which has operated since the completion of the experimental development in 1968 (see Fig 2-9). A broad programme of comprehensive investigations was an integral part of the project since it was to be the forerunner of such major tidal developments as 10,000 MW at Tugur Bay on the Sea of Okhotsk and 15,000 MW at Mezen Bay on the White Sea. [Bernshtein, 1986].

## Annapolis, Nova Scotia, Canada

A number of promising sites for tidal power development on the Bay of Fundy have been under consideration for many years. At one of these, the existence of a causeway, constructed in the early 1960's for marshland reclamation, has allowed use of the valley upstream as the tidal basin for a 20 MW tidal power development, commissioned in 1984. The progressive step here was to employ the Straflo turbo-generator design for tidal power. The unit size, with its 7.6m diameter runner, provides a true prototype for future large-scale applications. With the exception of a design weakness in the field windings of the generator which required modification, the plant has proved reliable with only modest maintenance requirements (see Fig 2-13). [Douma 1984].

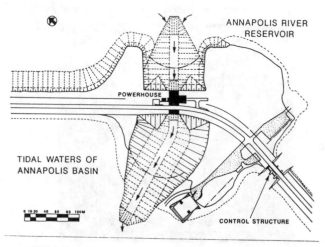

Fig 2-13. Annapolis Tidal Power Plant, Bay of Fundy

Fig 2-14.  Jiangxia Tidal Power Station.

### Jiangxia, Zhejiang Province, PRC

A number of multi-purpose projects along the coast of China have incorporated mini tidal power plants.  In 1980, a rockfill dam across Jiangxia inlet, built for reclamation purposes and aqua culture, was altered to incorporate a tidal power plant which will now have an ultimate installation of 3.9 MW to be provided by five bulb-type and one "straflo-type" turbo-generators [SHP News, 1985].  The mean tidal range is 5m and the bulb units are capable of two-way generation.

### Recent Investigations

During the 1980's, several fairly extensive investigations or reviews of previous studies were undertaken to examine the feasibility of exploiting the energy of the tides at several locations around the world where coastal configurations and tidal ranges offer promising potential.

### Cook Inlet, Alaska, USA

A preliminary assessment of Cook Inlet tidal power potential and characteristics was completed in 1981 [Acres].  Of the possible sites, the two most attractive are in the upper reaches of the Inlet across Turnagain Arm and Knik Arm.  A plant at Eagle Bay in the latter embayment could support a capacity of 1440MW.  The next step involves a determination of the most economic capacity which truly matches system demand.

## Garolim Bay, Korea

In view of the volatility of the world
energy situation and reliance on
imported oil, Korea undertook
renewed examination of the Garolim
tidal power project. [KORDI, 1986.]
Recent studies have led to selection of
20 x 20MW generating units as
optimum but the project's economic
feasibility was not fully justified. An
update in 1988 [LEE et al 1989]
demonstrated a slight improvement in
the economics of the project.
Development was concluded to be
uneconomic at that point in time.

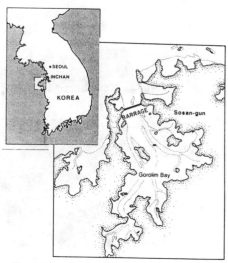

(Basin Area 50 Sq.Km Length 20Km

   Length of barrage 20Km

   Mean tidal range 4.9M.)

Fig 2-15. Possible barrage sites on the
west coast of Korea.

## Gulf of Kachchh, India

The need for diversity in use of indigenous fuel sources in India has encouraged
investigation of the 600 MW, single effect development on the north western coast
of India in Gujarat Province. The tides there have a mean range of 5.3 metres.
The proposed barrage alignment which would create a tidal basin of 278 square
kilometres would be formed by powerhouse and sluiceway floated-in caissons and
a rockfill section; a low embankment of about 4 metres in height and over 20Km
in length would also be required across the Sethsaida Bet to isolate the basin.

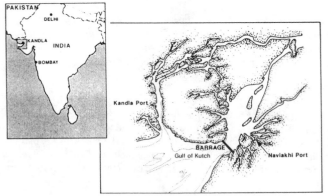

Fig 2-16  Proposed tidal power project in the Gulf of Kachchh.

## Bay of Fundy, Canada

The tidal power potential of the Bay of Fundy on Canada's Atlantic coast has been recognised and studied on repeated occasions during this century. Very high mean tidal ranges occur in the Cumberland and Mines Basins of 10.5 metres and 12.4 metres, respectively. The 20MW prototype development at Annapolis Basin has been noted (page 20). Development sites at A8 (Cumberland) could develop 1428MW and at B9 (Minas Basin) 5338MW (Fig 2-17). The results of recent reviews (Tidal Power Corporation, 1982) and studies (Alcan, 1985) have shown that the exploitation of tidal energy at both Sites A8 and B9 would be economic, although their outputs would not be competitive with thermal plants under present interest rates and treatment of social costs.

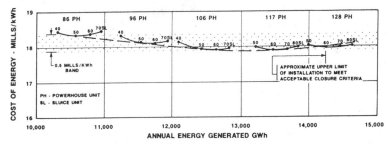

Fig 2-17.  Optimisation of installed capacity, Site B9 [Tidal Power Review Board 1977]

## Severn Estuary, United Kingdom

The Severn Estuary has been under consideration as a potential tidal recovery site for most of this century. Extensive studies, which more recently have concentrated on environmental issues, have identified a preferred scheme with 216 x 40 MW turbine-generator units (with 9.0 metre diameter turbines) operating in the fixed speed, ebb generation mode with flood pumping to increase energy capture and obtain maximum benefit from the tariff structure on the UK power system. The tidal power plant would have an installed capacity of 8640 MW sending out about 17 TWh per annum from a barrage across the Severn Estuary.

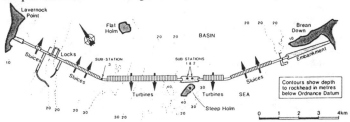

Fig 2-18  Layout of barrage for preferred Severn Tidal Power Scheme.

Mersey Barrage, United Kingdom, and other Medium Capacity Sites

Several estuaries in UK, including the Mersey on the West coast, have characteristics which make them attractive locations for the recovery of tidal power. The barrage length at the Mersey would be about 1.2Km accommodating 28 generating units of 25MW capacity and twenty 12m square sluices. Net annual output would be about 1500GWhr. The Mersey estuary is in a heavily developed region and is used by commercial shipping. Technical, economic, environmental and regional development aspects of the proposed project are being examined actively under joint government/private sector funding arrangements [Reilly and Jones, 1989].

The west coast of UK is favoured with other potential sites which have also been examined, some under similar funding approaches [ETSU, 1989]. The most promising are at Conway in North Wales (31MW and 60GWhr/year with mean tidal range of 5.2m) and at the River Wyre near Fleetwood in Lancashire. (47MW, 90GWh/year with mean tidal range 5.8m.)

### Modelling Tidal Plant Effects

A very major problem in any relatively large tidal power development is the influence of the barrage and its operation on the tide itself. Not only may the barrage and installation change the resonance aspects of the estuary but they may also change significantly the local energy phenomena which could entail significant impacts on the present regime. Such effects must be known from the start of the studies since the economics of the tidal plant as well as navigation in harbours and access channels and foreshore interests in the vicinity and even hundreds of kilometres away depend upon an accurate determination of the changes.

In pursuing this aspect of the La Rance development in the 1950's during the study and design stages, Electricite de France used a series of hydraulic models, including revolving models to take into account the Coriolis acceleration. At that time, this was the only modelling technique available and, of course, it was not possible to construct such models to the ocean boundary where the tidal amplitude would be unaffected. Usually for any important hydroelectric development, a hydraulic model is constructed to investigate complicated boundary and flow patterns etc., during and after construction.

The spectacular growth in numerical modelling techniques during the past 20 years with the advances in computer science has practically eliminated the use of analog models for tidal estuary investigations. Mathematical models are now the prime tool for the determination of the "far-field" effects of tidal power developments.

The continental shelf is now generally accepted as the boundary of the mathematical model. In the case of the Bay of Fundy, for example, the initial comprehensive studies carried out in 1966-1969 [ATPPB, 1969] placed the boundary of its mathematical model at the mouth of the Bay of Fundy. However, for the later reassessment studies [Tidal Power Review Board, 1977] the mathematical model limits to determine the effects of plant operation were extended to the edge of the continental shelf. The computed effects by the previous study were shown to be in error since it became obvious that the Bay of Fundy and Gulf of Maine acted as a single system. The 1977 model results were verified by a number of offshore gauging stations placed along the edge of the continental shelf at depths varying from 150m to 250m; measurement of the various tide and current components were taken over two to three month periods.

The advanced techniques in both hydraulic and mathematical modelling can now be linked to create a "hybrid" model. Thus, a hydraulic model of the particular area of a barrage location could be constructed at such a scale factor that distortions would be small and two-dimensional flow patterns meaningful could be interfaced dynamically with a numerical model which would be extended as far afield as necessary to ensure effective dissipation of reflections at its outer boundary. As implied, information would be exchanged between there two models almost instantaneously so that both would run together as one large model of the whole area. The hydraulic model would handle all those non-linear problems for which it was designed and the numerical model would look after the dissipation of reflections which it can manage quite well, and so eliminate the "boundary" problem of the physical model [Funke, 1981].

### Environmental Aspects

There can be no question that a tidal power development has substantial environmental impact both on the region surrounding the estuary closed off by the barrage and to the areas seaward of the barrage. The volumes of water transferred through the barrage are large and patterns of flow and velocity will change from those existing under natural conditions. Sediment movements and the contours of its deposition on the estuarine flow will also change and ultimately new land forms will be established.

The construction of a barrage which alters the regime of tidal movements will have an immediate and direct effect on the environment of the intertidal zone and upon the animals and plant life there. In general, there has been insufficient basic environmental study of estuaries and often as not this work has to proceed simultaneously with impact assessment of proposed development [Gordon and Longhurst, 1979]. Fortunately, however, increased attention is being applied, largely by national research institutes, to careful examination of the conditions existing in major (developable) estuaries and of the potential for management of these in a manner which will mitigate detrimental efforts while taking full advantage of positive benefits of environmental and regional effects [Severn Barrage: 1989]. In the previous section on Modelling it is noted that advances in techniques now allow for more confident prediction of the macro effects of projects on the tidal regime itself.

The potential for sediment movement, effects on water quality, impact on wading birds and other issues all need to be considered as part of site specific environmental impact assessments. Tidal level increases and fish mortality have been cited as two impacts which could yet impede development of sites with major potential, such as the Bay of Fundy [Baker, 1988].

| TIDAL PLANT FACILITIES AND CONSTRUCTION ACTIVITIES<br><br>* Indicates potential for interaction between environmental element and plant component | TOPOGRAPHY / BATHYMETRY | MINERAL RESOURCES | SOIL | DRAINAGE & SURFACE RUNOFF | GROUND WATER | TIDAL REGIME | FRESH / SALT WATER INTERFACE | PHYSICAL OCEANOGRAPHY | SEDIMENTATION / EROSION | WATER CHEMISTRY | AQUATIC ECOSYSTEMS | MIGRATING AQUATIC SPECIES | BENTHIC ECOSYSTEMS | INTERTIDAL ZONE PRODUCTIVITY | WETLAND VEGETATION | WETLAND WILDLIFE HABITAT | UPLAND VEGETATION | UPLAND WILDLIFE HABITAT | CLIMATE | AIR QUALITY | NOISE AND VIBRATION | ICE FORMATION | SURFACE WATER HYDROLOGY |
|---|---|---|---|---|---|---|---|---|---|---|---|---|---|---|---|---|---|---|---|---|---|---|---|
| **CONSTRUCTION ACTIVITIES** | | | | | | | | | | | | | | | | | | | | | | | |
| *SITE DEVELOPMENT - LAND BASED* | | | | | | | | | | | | | | | | | | | | | | | |
| CLEARING, GRADING, SURFACE EXCAVATION, BUILDING STRUCTURES, MATERIAL STORAGE | • | | • | • | • | | | | • | • | • | | | | • | • | • | • | | • | • | | • |
| ROAD, RAIL SPUR CONSTRUCTION | • | | • | • | • | | | | • | • | | | | | | | • | • | | • | • | | |
| EXCAVATION FOR ABUTMENTS | • | | • | • | • | | | | • | • | | | | • | • | • | • | • | | • | • | | |
| MATERIAL PLACEMENT | • | | • | • | • | | | | • | • | | | | • | • | • | • | • | | • | • | | |
| OPERATE LAND BASED MARINE EQUIPMENT | • | | • | • | • | | • | | | • | | | | • | • | • | • | • | | | • | • | |
| WORKER FACILITIES AND USE | • | | • | • | • | | | | • | | | | | | | • | • | • | | | | | |
| *SITE DEVELOPMENT - MARINE* | | | | | | | | | | | | | | | | | | | | | | | |
| PILE DRIVING | • | | • | | | | • | • | • | • | • | • | • | • | • | • | | | | • | • | | |
| INTERTIDAL CONSTRUCTION ZONE | • | | • | • | | | • | • | • | • | • | • | • | • | • | • | • | • | | • | • | | |
| DREDGING | • | | • | | | | • | • | • | • | • | • | • | • | • | • | | | | • | • | | |
| MATTRESS / DIKE PLACEMENT | • | | • | | | | • | • | • | • | • | • | • | • | • | • | | | | | • | | |
| TUG AND BARGE OPERATION | • | | | | | | | • | • | • | • | | | | | | | | | • | • | • | |
| CAISSON STORAGE AND TRANSPORT | • | | | | | | • | • | | • | • | • | • | | | | | | | • | • | | |
| CAISSON INSTALLATION | • | | | | | | • | • | | • | • | • | • | | | | | | | • | • | | |
| STATIONARY MARINE EQUIPMENT | • | | | | | | • | • | • | • | • | • | • | | | | | | | • | • | | |
| MECHANICAL / ELECTRICAL EQUIPMENT INSTALLATION | | | | | | | | | | | | | | | | | | | | • | • | | |
| *SITE ACCESSIBILITY* | | | | | | | | | | | | | | | | | | | | | | | |
| ROAD, RAIL TRANSPORT OF PERSONNEL, MATERIALS OR EQUIPMENT | | | • | | | | | | | | | | | | | | • | • | | • | • | | |
| MARINE TRANSPORT OF PERSONNEL, MATERIALS OR EQUIPMENT | | | | | | | • | | • | • | | | | • | • | • | | | | • | • | | |
| *REMOTE CONSTRUCTION FACILITIES* | | | | | | | | | | | | | | | | | | | | | | | |
| CONSTRUCTION MATERIAL SOURCE AREAS | • | • | • | • | • | | | | | | | | | | | | • | • | | • | • | | |
| DREDGE DISPOSAL SITES - UPLAND | • | | • | • | • | | | | | | | | | | | | • | • | | • | • | | |
| DREDGE DISPOSAL SITES - MARINE | • | | | | | | • | • | • | • | • | • | • | | | | | | | | | | |
| **OPERATION OF PERMANENT FACILITIES** | | | | | | | | | | | | | | | | | | | | | | | |
| ACCESS AND CLOSURE DIKE (PRESENCE) | • | | | • | • | • | • | • | • | • | • | • | • | • | • | • | • | • | | | | • | • |
| PHYSICAL ESTUARY BARRIER | • | | | • | • | • | • | • | • | • | • | • | • | • | • | • | | | • | | | • | • |
| POWERHOUSE AND SLUICEWAY (PRESENCE) | | | | • | • | • | • | • | • | | • | • | | | | | | | | • | • | • | |
| TURBINE OPERATION | | | | • | • | • | • | • | • | | • | | | | | | | | | • | • | • | |
| SLUICEWAY OPERATION | • | | | • | • | • | • | • | • | | • | | | | | | | | | • | • | • | |
| POWER FACILITIES (PRESENCE) | • | | • | | | | | | | | | | | | | | • | • | | | • | | |
| SWITCHYARD OPERATION | | | | | | | | | | | | | | | | | | | | | • | | |
| DRYDOCK AND DOCK FACILITIES (PRESENCE) | • | | • | | | | | • | • | • | | | • | • | • | • | | | • | | • | • | |
| LONG-TERM OPERATION | | | | | | | | • | • | | | | • | | • | • | | | • | | • | • | |
| IMPOUNDMENT (PRESENCE) | • | | • | • | • | • | • | • | • | • | | | • | • | • | • | | | • | | | • | • |
| WATER LEVEL FLUCTUATION | • | | • | • | • | • | • | • | • | • | | | • | • | • | • | | | • | | | • | • |
| LOCKS (PRESENCE) | | | | | | | | | | • | | | | | | | | | | | | | • |
| OPERATION | | | | | | • | • | • | | • | • | | | | | | | | | | • | • | |
| SITE ACCESS (PRESENCE) | • | | • | | | | | | | | | | • | • | • | • | • | | | | | | • |
| ROAD, RAIL SPUR USE | | | | | | | | | | | | | | • | • | | • | • | | • | • | | |
| MARINE USE | | | | | | • | | • | • | • | | | | • | • | • | | | | • | • | • | |
| WORKER FACILITIES (PRESENCE) | • | | • | | | | | | • | | | | | | • | • | | | | | • | | |
| USE OF WORKERS | | | | | • | | | | | • | | | | | | | | | | | • | | |

Fig 2-19 Matrix of environmental effects of tidal power development

Environmental impacts will arise initially from construction activities, no doubt increasing with effect as these proceed, and later from the operation of the permanent facilities Fig 2-19 shows the elements of the environment affected by the construction and operation activities involved [Acres American, 1981]. It will be apparent from this matrix that wide ranging multi-faceted studies are essential. They are being performed or contemplated for major tidal power recovery projects such as Bay of Fundy (Canada) or Severn Barrage (UK). Uncertainties are being reduced as work proceeds. If anything, confidence is increasing that mitigation costs and demonstrable benefits would have a net effect which would not erode the potential viability of a basically economic scheme.

Tidal power development can have positive environmental benefits. Maintenance of water level at a higher elevation with reduced fluctuation can ultimately bring advantages to coastal communities and life. Above all, as a renewable non fuel burning energy source, tidal power makes its contribution to reducing noxious emissions and $CO_2$ effects on the global scale.

Conclusions

Of all the methods of extracting energy from the oceans of the globe, tidal power probably enjoys the greatest level of proof of practicability from small to mega project scale. This chapter has concentrated on medium to large sized developments by recording achievements in actual construction and future plans based on the state of the art existing in the 1990 decade.

It is likely that economic considerations (as outlined in chapter 4) will have greater bearing on implementation of future tidal power development than will new technological progress (the topic of Chapter 5). The relatively high initial capital cost and the "money cost" implications of the long construction schedule combine to load tidal power with high front and financing charges. The "window of opportunity" for tidal power will be one coinciding with an era of low discount rate and supportive national fiscal policies. This is extremely difficult to forecast or, perhaps, contemplate.

The difficulty of making accurate predictions of the effect on the environment has, so far, been a deterrent to further tidal power development. However, growing concerns over potentially damaging global effects of alternative modes of electricity generation (even when "cleaned-up") may yet force a new philosophical attitude regarding investment to sustain a habitable ecology for mankind.

*References*

"*Feasibility of Tidal Power Development in the Bay of Fundy*", Atlantic Tidal Power Programming Board, Ottawa. October 1969, 5 vols.

"*Reassessment of Fundy Tidal Power*", Bay of Fundy Tidal Power Review Board, Ottawa November 1977.

Gibrat, R., "*L'Energie des Mares*", Presses Universitaires de France Paris, 1966.

Bernshtein, L.B.,         "*Tidal Energy for Electric Power Plants*", translated from Russian.
                          "*Kislyaguba*" A small station generating great expectations".
                          "*Tidal Power Engineering in the USSR*".    Water Power and Dam
                          Construction, March 1986, pp 37-40

Seoni, R.M. , "*Major Electrical Equipment proposed for Tidal Power Plants in the Bay of Fundy*", presented at the IEEE Power Engineering Society meeting Mexico City, July 1977.

Duff G.F.D, "*Tidal Resonance in the Bay of Fundy*", Journal of Fisheries Research Board, 1969.

Greenberg, D.A., "*The Effects of Tidal Power Development on the Physical Oceanography of the Bay of Fundy and the Gulf of Maine*", Canadian Technical Report of Fisheries and Aquatic Science, 1984.

Holler and Miller, "*Bulb and Straflo Turbines for Low Head Power Stations*", Escher Wyss News, Zurich, 1977.

Hadley, B. and Lindestrom, L.E. "*The Sidney A Murray Jr. Hydro Power Project*" Inst. Mech Eng. Seminar April, 1990.

Tidal Power from the Severn Estuary, Department of Energy, London, U.K., Paper 46.

Severn Barrage Project, General Report 1989, Department of Energy, London, U.K., Paper 57.

Cotillon J. "*La Rance Tidal Power Station, Review and Comments*", Proc. Colston Symposium 30, Tidal Power and Estuary Management, April 1978, Univ. of Bristol, Scientechnica, pp 49-66.

Braikevitch M., "*The Straightflow Turbine with Rim Generator*", Proc. World Power Conf 1970, World Energy Congress.

Douma, A. "*The Annapolis Tidal Power Project*", CEA Hydroelectric Power Section, Fall meeting, 1984.

Alcan (1985) "*Fundy Tidal Power Stage 1*" Report of Alcan Aluminium Limited to Tidal Power Corporation, Halifax. December 1985.

Daborn, G.R., (ed) "*Fundy Tidal Power and the Environment*", Proc. Workshop on Environmental Implications of Fundy Tidal Power, Acadia University Institute, Wolfville, Nova Scotia, Canada, November, 1976.

Gray T.J. and Gashus, O.K. (eds) "*Tidal Power*", Proc. of Int. Conf. on the Utilisation of Tidal Power held at Nova Scotia Tech. College, Halifax. May 1970, Plenum Press, New York - London, 1972

Tidal Power Corporation (1982), "*Tidal Power Update '82*", Halifax, N.S., 1982.

Funke, E.R. "*Hybrid Modelling as Applied to the Bay of Fundy*", Proceedings 8th Canadian Congress *of Applied Mechanics, Moncton, June 1981.*

*Reilly N. and B.I. Jones, "Progress on Civil Engineering and Planning of a Mersey Tidal Project", Third Conference on Tidal Power, London 1989.*

*E.T.S.U, "Taking Power from Water", Department of Energy, UK, 1989.*

*Acres American Inc. "Preliminary Assessment of Cook Inlet Tidal Power Phase I Report". Office of the Governor of Alaska. Columbia, Md. Sept. 1981.*

*Clark, R.H. "Tidal Power and Canada - A Review", Report to Department of Energy, Mines and Resources, Ottawa, Canada, 1987.*

*Dean, R.G. Tides and Harmonic Analysis, Chapter 4. In: Estuary and coastline Hydrodynamics (A.T. Ippen, ed.). Engineering Societies' Monographs, McGraw-Hill, New York, 1966.*

*Dawson, W.B. The Tides and Tidal Streams. Department of Naval Services, Ottawa, 1920.*

*Ippen, A.T. and Harlemann, D.R.F. Tidal Dynamics in Estuaries, Chapter 10: Estuary and Coastline Hydrodynamics (A.T. Ippen, ed.). Engineering Societies' Monographs, McGraw-Hill, New York, 1966.*

*S H P News "Jiangxia Tidal Power Station". Newsletter of the Asia-Pacific Regional Network for Small Hydro Power, No - 2, 1985, issued by the Hangzou Regional Centre for Small Hydro Power.*

*KORDI, "Korea Tidal Power Study - 1986, Vol II". Korea Electric Power Corporation.*

*Lee, D.S., Yum, K.D., Song, W.O. "Review of Garolim Tidal Power Project". Proc. International Conf. on Ocean Energy Recovery, Honolulu, 1989.*

### Bibliography

*Song, W.O., (ed) "Bibliography on Tidal Power 1483-1984", Korea Ocean Research Development Institute, Seoul 1984.*

*Moyse, C.M. "Bay of Fundy Environmental and Tidal Power Bibliography", Bedford Institute of Oceanography, Dartmouth, N.S. Fisheries and Marine Services Technical Report #822, November, 1978.*

# 3:   ECONOMICS OF TIDAL POWER

## T.L. Shaw

## 3.1  Introduction

Reference was made in Chapter 2 to the large number of possible
tidal power schemes which exist in countries all over the world.
The reasons why this resource remains almost totally untapped are
essentially economic.  To explain this and contemplate the future
requires that we address two questions:

-   if there is some deep-rooted reason why tidal power is
    inevitably relatively uneconomic compared with the more
    popular energy sources, what is it that sustains worldwide
    interest in its development?

-   what circumstances will need to change and by how much
    before the most promising schemes become economic?

The answer to these questions relates to the factors which
determine the economics of tidal power.  We also need to know
whether interest in tidal power is either commercially genuine
or is sustained simply because the technology is sufficiently
straightforward to make it an easy option to promote.

With this information we can contemplate the future, perhaps not
a demanding task but one which merits consideration because of
the size and availability of this energy resource.

## 3.2  Economic issues - Costs

As with any other project, the cost of tidal power schemes
comprises repayment of the initial loan with interest according
to borrowing rate and project construction period.  Annual
charges for operating, maintaining and repairing the facility and
providing services (transmission lines, ground rental, rates and
other charges) must also be covered.

The zero fuel cost and unusually long life of barrages and their
generating equipment are in their economic favour.  These help
to offset the usually adverse effect of high unit capital costs.
(This unit cost is often quoted as the total cost divided by the
installed generating capacity;  however, to be fully instructive
it must also recognise  plant load factor).

------------------------------

1. Consulting Engineer, Shawater Ltd., Ston Easton, Bath, U.K.

Fig. 3.1 shows the relative significance of capital, interest and other costs on the unit cost of producing electricity from the tides according to the required rate of return on investment. The dominant influence of the capital element is obvious; it accounts for about 85-90% of the total energy cost, this % increasing with the required rate of return.

The actual capital cost of any barrage (hence its unit cost) depends substantially on site conditions, e.g. geology and water depths. These determine foundation conditions, the dredging task and the size of structures. Tide range, basin area and barrage length are also key factors in determining the design of any scheme and its energy potential. The preferred construction method would be selected with reference to cost and other factors, balancing the financial disadvantage of an extended construction period against the greater facility cost of a shorter period.

The revenue potential of tidal power barrages is considered in Section 3.3. Unlike conventional thermal power stations, the method of operation of a barrage at any site significantly affects its economic performance due to interplay between generating capacity, its design and how it is used. Although conceptually simple, the optimum procedures for harnessing tidal energy remain to be worked out.

However, considerable progress with this has been made during the last ten years. The work has built on the more elementary information produced most notably by Gibrat in the 1950's when the Rance Barrage in Brittany was designed (Gibrat, 1966). Data from a series of studies of the Severn Barrage illustrates this recent progress. Table 3.1 shows how estimates of the average annual electricity output, installed capacity and capital cost of basically the same project changed over a ten year period.

TABLE 3.1

| Date of Cost Estimate | Installed Capacity (GW) | Ann. Energy Output (TWh) | Basic Capital Cost * £M |
|---|---|---|---|
| 1981 (S.B.C.) | 7200 | 12.9 | 5660 |
| 1986 (S.T.P.G.) | 7200 | 14.4 | 5660 |
| 1989 (D.En et al) | 8640 | 17.0 | 8280 |

(* Excluding power transmission)

These data show that the ratio of capital cost to annual energy output varied little in real terms over the period covered. This illustrates the advances made in optimising the performance of the scheme. The fact that the effects of inflation are included in these costs and that these rose much less quickly than the effects of inflation confirm the technical advances made with this project, both with respect of the capital required and the energy supplied.

Based on this experience plus indications as to how further improvements to this scheme may be made, an annual energy output of 20 TWh now looks possible without significant increase in 1989 prices.

Unlike thermal power stations, most of the cost of tidal power schemes is in the civil works. The division is typically 70:30, the reverse of thermal stations. This emphasises the need to look to the civil works to improve the economic case. Although Table 3.1 shows how much progress has been made in this area, radically different forms of acceptable design and/or construction may be needed to sustain it.

Tidal power barrages are commonly regarded as sources of energy alone, especially when they are designated in this way. It is easy to forget that, unlike most other energy forms, barrages also provide 'amenity' functions, e.g. a road crossing, recreation opportunities and enhanced riparian land values. Since the case for amenity barrages hinges on their 'non-energy' merits including their political appeal and the employment generated, these factors must be taken into account in their valuation.

It has been said by developers of these projects that while they are straightforward to promote according to the projected sale value of electricity produced, the benefits of their 'non-energy' features are comparatively uncertain and hence should be treated as a windfall. On this argument, amenity barrages have no real value, yet they have been and are increasingly promoted, constructed and operated with financial success.

An important factor here is the extent to which investors in an energy barrage can expect to benefit from its particularly diverse advantages. Generalised answers are difficult because each scheme is unique, its merits largely depending on its location relative to population centres. For most barrages, economic viability hinges on the cost of energy production. The Rance Barrage confirms the importance of non-energy benefits, an aspect applicable to many schemes proposed for the UK and elsewhere.

### 3.3 Economic Issues - Revenue

It is usually straightforward to make a first estimate of the annual energy potential of a tidal power site, though to refine and optimise this is less easy (as noted in Section 3.2). Given a reasonably detailed hydrographic map and a forecast of annual tide ranges close to the proposed barrage site, and making allowance for the effect of the operating barrage on tide regime, the energy potential per tide and hence the annual output may be determined.

This estimate must be based on an assumed generating capacity. This capacity will recognise any constraints imposed by geology and/or water depths on the plant which can be accommodated.

The financial value of the energy produced will depend on the regime into which it is sold and how that regime may change over the economic lifetime of the project. There are many reasons why this value is unlikely to fall in real terms. Environmental and supply constraints, the general prominence of green issues and the ability of politicians to influence international energy markets are reasons why energy prices are likely to move ahead of inflation, though these are not unrelated because of the influence of energy prices on industrial costs.

In this situation, investors will be comfortable with financial calculations based on constant annual revenue in real terms; inflation is unlikely to erode their position. Any extra income, which in practice is likely to be substantial, will tend to be treated as an added but financially less-quantifiable benefit.

There are many misconceptions about the value of the tides as an energy source. Conventional wisdom is based on the notion that this more-or-less predictable input may be harnessed when available by simple machines to give intermittent slugs of electrical supply to networks which ought to be fed regularly because their consumers have regular demands.

However, this perception of the availability of tidal power and how networks operate is misleading. First, consumer demand is not regular. Second, the operation of tidal power plant is not rigidly constrained by tidal phase and range, especially when the machinery incorporates the features in use at the Rance Barrage. The lessons which have come from that scheme over 25 years should ensure that its design remains compatible with the efficient operation of the electrical network it supplies. Some details of how this is achieved and why the lessons are generally relevant to tidal power development are given in S.T.P.G. (1986).

Uncertainty over the sources of future energy supplies and their costs emphasise the need to incorporate flexibility into the design of tidal power schemes. Furthermore this adds value at nominal additional cost, raising load factors by ensuring greater power output over longer periods during the mid and lower range tides.

Reference was made in Section 3.2 to non-energy aspects of tidal power barrages including the possible difficulty, depending on scheme ownerships, of attributing any part of the revenue from that sector to investors in the barrage. It is unlikely that a significant part of that additional revenue will be realised without capital investment beyond that in the power station, e.g. a public highway rather than a service road across the barrage, marinas and hence additional locks in the barrage to take advantage of the reduced tide range for recreation, and a visitor centre and other attractions for tourists.

The ownership of these 'non-energy' facilities, also the land associated with them and other developments attracted to the area by the presence of the barrage are basic to determining how revenue from these facilities would be attributed. Any division between public and private sector investment will depend on the political regime in which each scheme is promoted; this needs to be considered on an individual basis and cannot usefully be discussed in abstract.

## 3.4   Environmental Considerations

In order to generate electricity, water is released from and subsequently refills the basin formed by the barrage. These operations are closely related to tidal motion and they cause a substantial rise and fall of water level in the basin, typically 40-50% of the range of the higher tides which occurred before barrage construction (see Chapter 2).

Depending on whether energy is generated on the ebb or flood tide, the average water level of this modified regime would respectively be higher or lower than the former mean level. The several clear advantages in adopting ebb rather than flood generation are explained in Chapter 2. The principal features of the modified tide regime are fundamental to determining the wider environmental effects of a barrage. One fact is immediately clear, namely that conditions in the basin would not resemble those of a reservoir. In fact, the array of water levels occurring in the basin would still be higher than those on most coastlines of the world because tidal power barrages are only likely to be viable where tide ranges are particularly high.

The dependence of the related environment on tide range means that the changes which any barrage will cause are likely to be significant in physical and ecological terms. However, actual changes which occur will not be the same for all schemes because each estuary has its distinctive features. Also, whether or not those changes are acceptable however this is measured (e.g. with reference to the environmental effects caused by the generation of similar amounts of electricity by other sources), is largely subjective and will be differently made according to perceptions of what is significant.

While it is not pertinent to debate subjective environmental judgement here, there is usually scope in the design, construction and operation of a tidal power barrage (as with any project) to moderate the implications when these are established and the required direction of change is stated. Some proponents of reactive response call this 'creative conservation', others refer to it as amelioration. Volume 4 of D.En et al (1989) gives considerable attention to this subject in the context of the Severn Barrage scheme; by carefully considering the environmental changes which <u>could</u> occur if certain decisions about that scheme were taken, and by focusing on how these depend on the details of design, construction and operation, the direction and scale of change has been shifted towards more acceptable solutions, without compromising the energy potential of the scheme. Furthermore, the developments reported are likely to be generally applicable though their application will differ from estuary to estuary because of the uniqueness factor referred to above.

The popular media tend to highlight sedimentation, water quality and shorebirds as subject areas adversely affected by barrages. Superficially the reasons for this are clear. Viewed more closely, however, the outcome is less obvious and some positive environmental advantages have been identified. Volume 4 of D.En. et al (1989) explains why this is so. The research information now available suggests that topics like the safe passage of fish may be of greater environmental importance than those more commonly identified, though more study is needed to establish the true position and how it can be made most acceptable. There is no substitute for site specific studies to establish the relevance of the rapidly growing body of basic information relevant to each estuary. Furthermore, no appraisal could be complete if it deals only with the environment affected by tide range; the geographically much wider consequences of change to regional infrastructure may in many respects be more influential and hence worthy of close study.

3.6 CONCLUSIONS

The factors which determine the economics of tidal power have been considered in this Chapter. There is no simple way of measuring commercial viability. Capital costs including interest payments stand out as the major financial penalty to be overcome by revenues, but neither the costs nor the benefits of energy sales are as yet presentable in generalised form; indeed this may not be a sensible objective because of the difficulty of expressing those factors which characterise and distinguish individual schemes.

It is also true that the basic information needed to optimise the energy value of any scheme according to the characteristics of the network supplied is not fully available, though work is continuing to provide this.

Another important shortfall in most economic assessments has been the lack of quantitative recognition given to issues other than power generation.  A growing body of evidence suggests that for those schemes located in areas in which public infrastructure either is or could readily be developed, the presence of a barrage should beneficially stimulate economic growth.  Indeed, the likeness of many potential regional developments closely resembles those expected of so-called amenity barrages, in which power generation does not feature.

While tidal power barrages cannot be considered as amenity barrages with the addition of power generating facilities,  some division of cost between the two functions has to be equitable for both interests to be taken into account.

The case sometimes made about the probable difficulty of connecting the non-energy benefits of this type of project to its capital financing is relative rather than real.  The necessary approach is foreign to the present energy sector whose product market is clear and relatively straightforward.  The 'amenity' business is altogether different but is no less commercial, real and potentially successful as a result.  Promoted objectively it will substantially improve the economics of many possible power schemes;  and not to do this would be to put schemes at risk, possibly losing an opportunity for the many interested parties to benefit.

In this respect, designers and promoters need to have regard for those likely conditions in the middle and longer terms which will affect project economics and determine financial credibility.


<u>Appendix</u>

Gibrat, R., 1966, L'Enérgie des Marèes, Presses Universitées de France

Severn Barrage Committee, 1981, Tidal Power from the Severn Estuary, UK HMSO Energy Paper No. 46

Severn Tidal Power Group, 1986, Tidal Power from the Severn, Thomas Telford Limited

UK Department of Energy, CEGB and STPG, 1989, The Severn Barrage Project, UK HMSO Energy Paper No. 57

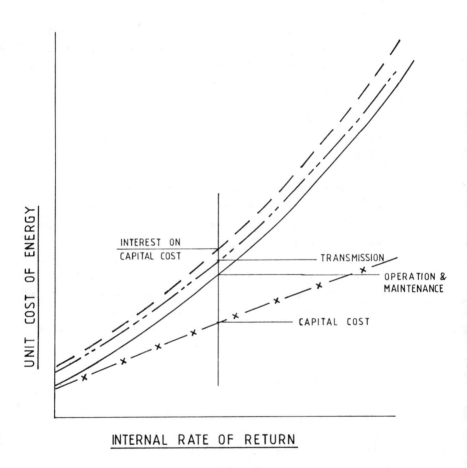

INTEREST ON
CAPITAL COST

TRANSMISSION

OPERATION &
MAINTENANCE

CAPITAL COST

UNIT COST OF ENERGY

INTERNAL RATE OF RETURN

FIG. 3.1

# 4: NEW TECHNOLOGY APPLICABLE TO TIDAL POWER

## G. C. Baker[1]

## 1. Introduction

The development of technology for modern tidal-electric plants began with the La Rance project in the early 1960's and now spans about three decades. During much of this period, sustained or sporadic searches for technical improvement have been conducted in several countries, including France, Russia, the United Kingdom and Canada. As a result, tidal technology is in some respects approaching maturity. Given these circumstances, what opportunities may exist for future improvements?

Surprisingly, the main difficulty in answering this question does not lie in paucity of ideas because tidal power has for at least the last century held a remarkable attraction for inventors. The problem is to decide which, if any, of the numerous alternatives may have sufficient technical and economic potential to warrant serious consideration. To provide a background for discussion, the character and objectives of technical investigations from La Rance to the present are briefly reviewed.

## 2. The Status of Present Technology

Early investigations had indicated that tidal power was not economically competitive with existing sources of electrical generation. Technical studies therefore tended to pursue two objectives; reduction of

---

[1]Consulting Engineer, 536 Main St., Kentville, Nova Scotia, Canada B4N 1L3.

capital cost per unit output, and increase in the value
of output by investing it with some degree of firm
capacity. In the last decade a third objective,
mitigation of environmental impacts, has assumed
increasing importance.

Because of its limited potential compared to
conventional energy resources, tidal power has not in
general enjoyed the level of funding necessary for
research and development of special-purpose machinery or
materials and tidal technology has therefore tended to
develop more by adaptation than by creation.

One obvious source of both machinery and concept
was river hydro, which, by the 1960's, had already
reached an advanced stage of technical development. Low
head turbines developed for use on rivers were adapted
for tidal use; for example, the Kaplan runner configured
as a bulb turbine at La Rance, or the straight-flow
turbine at Annapolis. The concept which has governed
the approach to capital cost reduction, insofar as
machinery is concerned, flows also from the river hydro
heritage. It has comprised site selection and other
means of achieving the highest possible operating heads,
together with utilization of turbines with the highest
practical specific speeds in order to reduce the cost of
rotating machinery.

Various trade-offs between efficiency and the
cost of turbine regulation have been explored, including
variable pitch versus fixed blading, single versus
double regulation and the use of variable speed turbines
coupled with rectification and inversion to achieve
output at power system frequency (1), (2), (3).

Other efforts to reduce capital cost have involved the adaptation of caisson construction (from ocean engineering technology) which was successfully demonstrated at Kislaya Guba; optimization of installed turbine, generator and sluice capacity, which tends to be site-specific and to vary with economic parameters; shortening of construction period and early barrage closure (3), (4); packaging of turbines and auxiliaries for installation in caissons to maximize shop work as opposed to field work and reduce installation time (5); use of steel rather than concrete for caissons (7); and use of a pumping strategy to increase head and output.

Tidal energy exists in only two forms; potential as manifested by tide height and kinetic as manifested by currents. Any attempt to extract such energy is therefore inevitably limited to the utilization of heads or flows. Efforts to date have focussed almost exclusively on exploitation of heads by means of conventional hydraulic turbines. The other option, sometimes proposed but usually considered to be economically inferior, is discussed in section 6 below.

Various types of plant, including single basin single effect, single basin double effect and linked basins, have also been investigated, from the point of view of both capital cost and output value. While any conclusions as to the relative virtues of these plant types must to some extent be site-specific, linked basins appear to be economically inferior, while the choice between single- and double-effect appears to be rather narrow and to depend to some extent on the characteristics and needs of the electrical system which

the tidal plant is intended to serve. More complex
configurations of linked basin schemes giving greater
output have been proposed, but their technical and
economic feasibility has not been determined. One of
these, the "delta" powerhouse (8), gives the option of
turbining from high basin to either low basin or sea.

The main avenue to improving the value of output
(and ease of assimilation by utility systems) is energy
storage. Types considered to date include pumped
storage and compressed air energy storage. Some
proposed variations of the latter method have not been
examined. Other possible storage media include hydrogen
and electric fields. Such possibilities are further
considered in section 4.

In recent years there has been growing concern
about the environmental impacts of human activities, and
particularly those resulting from energy production.
This has led to increased emphasis on the environmental
aspects of tidal plant design and increasingly stringent
requirements for project approval by environmental
authorities. When La Rance was built, there was
apparently little interest in such matters and
environmental studies were not undertaken until some
years after the station had entered service. There has
been no indication of any serious long-term impacts (9).
By contrast, an environmental assessment and more than
two dozen subsidiary studies on specific aspects were
necessary in order to satisfy the requirements for
approval of the Annapolis plant built about 20 years
later. The general nature of the impacts to be expected
from tidal plants and some technological requirements
for dealing with environmental problems are discussed in
section 5.

3.  Cost Reduction Possibilities

     For propeller turbines as for other types modern
design approaches have resulted in an upward trend for
both specific speed and efficiency.  For bulb turbines,
Schweiger and Gregori (10) found that units set between
1976 and 1985 had an increase of 25.8% in specific speed
over units set in the previous decade.  The explanation
appears to be that manufacturers had used part of the
loss reduction due to the straight conical draft tube to
increase throat velocity, runner speed and losses in
that part of the machine, resulting in an increase of
specific speed and some net reduction in total losses
(11).  The possibilities are therefore limited and
further increases in specific speed would probably
entail the penalty of higher losses.  Whether present
designs represent the best trade-off for tidal power
applications is not known, but the potential for
improvement in project economics through this avenue is
probably insignificant.

     Powerhouse caisson costs and the time and cost
entailed in installation of machinery are both affected
by turbine type.  Settings for bulb turbines, capable of
being placed or removed as complete units, have been
developed in connection with designs for the Severn
tidal power project (5).  Another type of turbine with
potential cost advantages is the pit turbine.  This type
is an evolutionary development of the bulb turbine,
replacing the bulb by a pit and using a speed-increasing
drive which results in a smaller and marginally more
efficient generator.  Other comparative advantages
include easier access and easier transfer of dynamic

stresses to the powerhouse structure. It lends itself
to unitized installation in caissons. At present gear
train capacity is said to be limited to about 40 MW.
Pit turbines may deserve consideration for tidal
applications, particularly if the unit capacity can be
increased to the levels deemed appropriate for larger
tidal power sites. Turbines of this type are proposed
for the Mersey project in the United Kingdom (25).

Another turbine type of potential interest uses a
bevel gear drive with the generator located above the
turbine. This minimizes water passage obstruction and
imposes much lower capacity limits.

The design of civil elements must reflect
geotechnical conditions, water depth, head and
overturning forces, closure requirements, material
availability and other circumstances which vary from
site to site. There can thus be no single best
approach. Designs prepared for various sites at the
feasibility level are probably not optimum, but also
probably not far from it. Nevertheless, there are many
unexplored or only partially explored possibilities for
innovation in design, use of materials and construction
methods.

Two techniques used for other purposes which
might be adapted to tidal power are the sand isle method
of creating stable offshore platforms and the pneumatic
devices developed for gating hydro spillways. In both
cases the range of applicability would be limited at
best, and there is no a priori reason to suppose that
such innovations would be economically advantageous.

Potential exists for benefit from additional RD and D in the area of civil works, not only in terms of further (and probably modest) cost reduction, but also through reduction of the perceived risks of tidal power technology.

## 4.  Energy Storage

The periodic nature of output from single basin tidal plants has been a matter of continuing concern to designers because such an intermittent source lacks firm capacity and its market value is correspondingly reduced. At present costs of conventional base load generation, capacity costs may amount to between 2 and 3 cents per KWh. The intermittent supply is of concern to potential utility customers for another reason:  the relatively sudden increase and decrease of tidal power input to the utility system might involve ramping thermal generation down and up at uncomfortable or even potentially damaging rates. For this reason, utilities could only absorb an amount of tidal power equal to a small fraction of the system load.

Storage of energy with release to the utility system as required is a conceptually simple solution. Methods studied have included pumped storage and compressed air energy storage (CAES) (12). Both are commercially proven. Pumped storage may entail between $1000 and $2000 per KW of capital cost depending on the sites available and would further entail in-and-out losses of about 25% of the input energy. CAES is estimated to be less costly at $700 to $800 per KW and requires inputs of about 0.75 KWh of tidal energy plus about 4000 Btu of gas turbine fuel for each KWh of output. CAES is thus economically more attractive than pumped storage but the fuel dependency may be an adverse factor.

A CAES facility can be configured in a variety of ways, of which only a few have been investigated, and none exhaustively. Compression could take place at the tidal plant, in which case the CAES plant would consist of storage and regeneration facilities only, generators at the tidal plant would be eliminated and pipeline costs would be incurred. Alternatively, electric power could be transmitted from the tidal plant to the CAES, in which case the latter might use common motor/generators, or separate motors and generators.

Other design considerations involve the amount of storage (whether sized to handle daily or weekly utility requirements, or also lunar fortnightly variations of input), the number of stages of compression and cooling and whether or not heat recovery is to be provided. A recent concept involves compression at the tidal plant and delivery of the compressed air to the gas turbine facility via a pipeline which also acts as the storage element of the system.

Tidal technology would benefit from further study of CAES, including the development of systematic procedures for optimizing design to meet the needs of tidal power plants.

Another option for energy storage, which may or may not be available in any particular case, involves the use of an existing hydro headpond as the storage medium. Operation entails reduced output from the river hydro plant while the tidal plant is operating and increased output when the tidal plant is idle. Providing sufficient transmission capacity is available, the costs might be of the order of $900 per KW for extra generating capacity at the river hydro plant. The tidal and hydro plants need not be directly interconnected so long as each can transmit to a common load and the combined firm capacity can be despatched from the river hydro plant. In-and-out losses would be small or non-existent. Where the required facilities are available, this is probably the best option for retiming.

Other energy storage systems may become economically attractive in future. Using hydrogen as the storage medium is now technically feasible and may become economically attractive if hydrogen comes into common use as a transportation fuel. Regenerating electricity from the hydrogen is at present too inefficient. Superconducting magnetic energy storage may become competitive if and when high temperature superconductors are developed to the point of availability for power cables.

## 5. Environmental Considerations

Tidal power plants tend to be physically large, at least in relation to their output, and inevitably alter the ebb and flow of the tides to some degree. The consequences for the natural environment are likely to be highly site-specific, but based on the results of extensive investigations in the Severn Estuary (13) and the Bay of Fundy (14), there are likely to be some common factors.

Changes in the location, extent and timing of tidal currents are likely to be of significant magnitude, particularly in the headpond area. The effect of such changes on the sediment regime are important from both engineering and environmental viewpoints. The engineer needs some assurance that the headpond will not become a sediment trap resulting in lost storage capacity. The ecologist needs some assurance that sediment build-up or erosion will not be so rapid as to wipe out benthic species such as clams. The ecologist further needs some prediction of suspended sediment concentrations in order to estimate the effect of the tidal development on biological productivity.

At present the means for satisfying such needs are notably inadequate.

Predictions of sediment behaviour tend to be of low accuracy because the processes of deposition and resuspension of sediment are not well understood. This situation was underlined by a recent discovery that energy levels needed to resuspend sediments, obtained by laboratory measurement to supply data for modelling purposes, were as much as 80 times lower than the levels obtained by making the same measurements on drying mud flats (15). Efforts to attain a better understanding of the underlying physical (or biological) factors are being undertaken (16). However, it is clear that at present the inability to deal accurately with sedimentation questions is a weak point in tidal technology.

The effect of tidal plants on fish and marine mammals is another area of general concern. Hydraulic turbines are known to impose some level of mortality on fish due to direct contact with turbine blades, cavitation, shear stresses, and sudden pressure drop at the turbine throat. The single passage mortality level depends on turbine size, flow rate, rotational speed, number of blades and operating condition as well as on fish length and species (17), (18). For anadromous fish making one spawning run or one run per year, the mortality level imposed by the large slow-speed runners typical of tidal power applications may be tolerable. However, turbines placed in estuaries may affect indigenous fish populations moving in and out with the tides and thus likely to be subject to multiple passages, in which case even a low single passage mortality rate might be unacceptable (19).

Fish mortality studies conducted in connection with the Annapolis plant have illustrated the difficulty of determining mortality with useful accuracy and one study (20) established a strong inference that generally accepted measurement techniques are seriously flawed.

Assuming that some measures to mitigate fishery impacts would usually be required, the obvious and practical solution is to provide separate fish passages. But this is only a partial solution, unless means are also provided to induce fish to use the safe passages provided. Experience with river hydro plants has indicated that the most dependable and effective diversion devices are travelling screens. However, devices of this type, at typical costs in excess of $10,000 per $m^3s^{-1}$ of turbine flow, would be prohibitively expensive if applied to the large flows of tidal plants. Also, such screens might be unsuitable or less effective if installed in an estuarial setting for other reasons as well. Tidal power technology should include economic and effective means of fish diversion, suitable for employment in an environment featuring two-way flows, salt water and marine fouling communities. While some trials have been made using behavioural devices, with encouraging results, including sound and floating screens, a general solution to this problem remains outstanding.

There is a continuing need for improved understanding of ecological systems and processes and for improved capacity for impact prediction, both to aid in the integration of engineering and environmental objectives and to reduce the perceived risks of tidal power projects.

6. Reciprocating Tidal Power Devices

Unlike conventional technology, these devices do not require high hydrostatic heads, and can be applied to locations with a tidal range as low as 2 m. They also do not employ high-speed water turbines, which greatly reduces their potential impact on fish.

The first of these is the Salford Transverse
Oscillator, developed in Scotland (26). This device
consists of a parallel, double-wall fence built across
the direction of flow in a tidal inlet. Water flowing
through gaps in the fence triggers controlled
oscillating motion of broad flaps, in a direction
transverse to the main flow (Figure 1). The flaps are
linked to the drive rod of a double-acting hydraulic
cylinder located onshore, which pumps high-pressure oil
to hydraulic motor/generators. Prototype designs have
been developed for two sites in the Outer Hebrides
Islands.

Several devices have been developed that produce
reciprocating air flow from low-head tidal and river
power. The first of these was developed by Dr.
Alexander Gorlov at Northeastern University in Boston,
Massachusetts (27). Dr. Gorlov's design also employs
watertight, reinforced architectural fabric for dam
construction, which greatly lowers capital cost
(Figure 2).

In England, Coventry Polytechnic has developed
two reciprocating air flow devices, based on wave energy
conversion technology (Figure 3). One is similar to Dr.
Gorlov's device, where sequenced opening of upstream and
downstream valves produces vertical, oscillating water
column motion inside a concrete caisson, which drives
air through a Wells turbine (28). A second device
developed at Coventry employs flexible bags of the type
used in the SEA Clam. Two such bags are placed in an
inclined duct. As water flows through the duct, it
alternately inflates and deflates the bags, to produce a

reciprocating air flow (27). In 1987, a 150 kWe flexible bag device was installed at Borrowash, on the River Derwent. This prototype operates on a hydrostatic head of 2.8 m (29).

7. Submerged Current Turbines

Like conventional tidal power plants, the use of reciprocating devices requires that a dam be built across a coastal estuary. As an alternative to dam construction, and in contrast to the usual strategy of maximizing head, it is quite possible to extract energy directly from tidal (or other) currents. Historical antecedents go back to the "Clarkson current generator" designed for an abortive attempt to harness the Bay of Fundy tides about 1915, or for that matter, back to the undershot water wheel.

Interest in current generators for possible use in Fundy was revived in recent years due to a combination of circumstances. Planned tidal power developments in Minas and Cumberland Basin were predicted to increase the tidal range in the Gulf of Maine, and to cause fish mortality. The barrages associated with conventional plant design seemed to be implicated and it was conjectured that the impacts might be avoided or mitigated through the use of current generators. At about the same time, the Canadian National Research Council was investigating the Darrieus turbine for wind power applications and was interested in assessing its prospects for employment as an underwater windmill.

Mathematical analysis indicated that the optimum location for current generators would be considerably to seaward of optimum barrage sites, and that energy recovery would be maximized by using runners of suitably low impedance (21). It further appeared that energy recovery would be too low unless the tidal stream could be confined to flow through the runners. This led to the tidal fence concept, in which a large part of the cross-sectional area of the tidal stream was filled with turbines, creating what may best be described as a highly permeable barrage or "tidal fence", across which would appear a head of 1 or 2 m at maximum flow.

Preliminary investigation of a tidal fence (22) spanning the narrow entrance to Minas Basin indicated that it would produce more energy than a barrage at site B9 with about the same effect on the Gulf of Maine tidal range. The inference, supported by other modelling, is that a suitably sited tidal fence could produce the same energy as B9 with less tidal perturbation.

Hopes for less fish mortality from a tidal fence do not appear to be justified, both because of the greater numbers of fish at risk in a more seaward location and because of the apparent impossibility of diverting fish to a safer route. A tidal fence would impose about the same mortality as a turbine runner due to direct strikes, less due to pressure drop and possibly less due to cavitation and shear.

No dependable economic conclusions can be drawn from the study. However, the Darrieus turbine and other technological aspects were years away from commercially proven status and the work was dropped at that point.

From an economic standpoint, the tidal fence suffers by contrast with the conventional approach due to the much lower energy density of the working fluid. The question is whether the extra costs due to the necessity of larger total prime mover throat area can be compensated by the avoided costs of barrage and sluices. The outcome is likely to depend on site characteristics. It is possible that under some circumstances the tidal fence would prove superior and that there might be some justification for further development of tidal technology in this area. Development of a suitable turbine and resolution of the problem of generator location are among the matters which would have to be addressed.

The most recent innovation in current turbine design is a unique horizontal-axis machine that incorporates an outer and inner turbine into a single unit (Figure 4). The blades of the outer turbine are hollow, permitting radial water flow due to the centrifugal force developed by the blades' rotation. A smaller turbine at the center of the outer turbine generates electric power from the axial flow before it enters the outer turbine blades. This inner turbine is exposed to a flow speed approximately five times greater than that of the free stream, minimizing the speed-increasing gear ratio required for synchronous electric power generation (30). Several prototypes have been tested, and a pre-commercial unit, 2.6 m in diameter, is now under construction, with a minimum rated output of 15 KWe.

8.  Is Small Beautiful?

        The burden of tidal plant capital costs can be
lightened through sharing if the civil works serve some
other purpose having economic or social value.  In the
case of large plants, with installed capacities in the
thousands of megawatts, the cost so shareable has tended
to be minor or negligible in comparison to the total.
However, for small plants the situation may be
different.  In China small tidal power plants have been
developed in combination with such other uses as water
level control or mariculture facilities, with the result
that economic justification has been obtained for energy
projects which would otherwise have been unattractive
(23).

        As a policy for tidal power development, such an
approach at first glance appears to forego the
advantages of scale which have led to designs with 4- or
5-figure installed capacities in Russia, the United
Kingdom and Canada.  However, feasibility studies of
sites of various size in the U. K. have shown a general
similarity of unit costs regardless of scale (7).  It is
possible that in an economic climate of high interest
rates, the advantages of scale are offset by longer
construction periods and the consequent burden of
accrued interest costs.  Under such circumstances, small
plants may actually be advantageous and might well be
included in development strategies.

        Among the unconventional ideas sometimes put
forward as the salvation of tidal power, a few may have
validity if applied to small-scale developments.  One
such is hydraulic compression, a technique frequently

used in the past to provide compressed air for mining
and industrial use.  Such a system is illustrated in
Figure 5.  The pressure developed depends on the depth
of the separation chamber and is relatively independent
of the hydraulic head.  However, for a given head, the
quantity of air compressed varies reciprocally with
pressure.  For small-scale applications, the stored
energy could be recovered by means of an air turbine.
For larger applications, a combustion turbine might be
more appropriate.  With adequate air storage, the output
would have firm capacity and other desirable
characteristics.

Adaptation of this technique to tidal power would
require some developmental work.  The first requirement
would be to establish the performance parameters
necessary for evaluation of feasibility.

Another relic of the past which might possibly
prove useful for very small-scale development of tidal
power is the hydraulic ram.  This device is capable of
raising water to a considerable elevation using the
kinetic energy available from modest stream flows well
within the range of tidal currents.  Water so pumped and
stored would provide firm power from a turbine operating
at relatively economic heads.  It is doubtful that
hydraulic rams could be scaled up sufficiently to supply
plants of significant size.  As in the case of hydraulic
compression, parametric information about such devices
would be necessary in order to make an informed judgment
as to their economic potential.

9.   Summary and Conclusions

     The main thrust of technical development has been
toward optimum design of single or double effect, single
basin tidal plants and for this purpose present
technology is relatively mature. Further improvements
will no doubt be achieved in the design of both overall
systems and individual components. However, the gains
in cost-effectiveness are likely to be modest.

     There is considerable scope for a better
understanding of how energy storage may be integrated
with tidal power to provide a high quality source of
generation for electrical systems. Neither experience
nor optimization methods are available to chart an
efficient course through the maze of systemic options.
This is an area which will probably attract more
attention in the future than in the past, particularly
if utility capacity costs increase to regain their
pre-OPEC relationship to energy costs.

     Better means of predicting sediment regimes and
efficient and cost-effective methods of fish diversion
are foreseen to be necessary to deal with environmental
aspects of tidal power design.

     The ultra-low-head approach is the only generic
alternative to the present direction of technical
development. It has some technical credibility but
suffers from the economic handicap imposed by extremely
low energy concentration. It might nevertheless be
advantageous under certain site conditions and if so
would broaden the potential for tidal power development.

Small tidal power plants may have advantages in an era of high interest rates, would provide more scope for variation of development technique, would lend themselves to incorporation in joint use projects and would also broaden the potential for tidal power development.

Subject to the uncertainty involved in any attempt to visualize the future, these are the areas in which tidal technology may expand. However, it should be borne in mind that tidal plants of conventional design could be economic now at favourable sites given real interest rates at historic levels (24). The comments in this chapter are therefore not intended to imply the need for abandonment of the conventional approach in favour of more speculative pursuits. The intent is merely to show where some broadening of focus might occur.

## REFERENCES

1.  Anon.; 1978; Reassessment of Fundy Tidal Power; Tidal Power Review Board; Ottawa, Canada.

2.  Anon.; 1981; Tidal Power from the Severn Estuary: Contractors' Reports Numbers STP 6, 7, 8 and 49; Department of Civil Engineering, University of Salford, Salford, England.

3.  Anon.; 1985; Fundy Tidal Power, Stage 1; Tidal Power Corporation, Halifax, Canada.

4.  Anon.; 1982; Fundy Tidal Power Update, 1982; Tidal Power Corporation; Halifax, Canada.

5.  Anon.; 1989; Severn Barrage Project, Detailed report - Volume IIIa, Civil Engineering; ETSU TID 4060-P3, Energy Technology Support Unit; Harwell, Oxfordshire, England.

6.  Anon.; 1989; Severn Barrage Project, Detailed report - Volume IV, Ecological studies, landscape and nature conservation, ETSU TID 4060-P4; Energy Technology Support Unit, Harwell, Oxfordshire, England.

7.  Anon.; 1989; Annual report of the tidal energy R & D programme 1988/89; Energy Technology Support Unit, Harwell, Oxfordshire, England.

8.  Baker, G. C., and E. van Walsum; 1983; Development Options for Fundy's Tidal Power Potential; Seventh Canadian National Energy Forum; Halifax, Nova Scotia; CANWEC, Ottawa, Canada.

9.  Banal, M., and A. Bichon; 1981; Tidal Energy in France: The Rance Estuary Tidal Power Station -- some results, after 15 years of operation; Paper K3, Second Symposium on Wave and Tidal Energy, September 1981, Cambridge, England.

10. Schweiger, F., and J. Gregori; 1988; Developments in the Design of Bulb Turbines; Water Power & Dam Construction; Sutton, Surrey, England; September, 1988.

11. Gordon, J. L.; paper in preparation.

12.  Anon.; 1988; Assesment of Retiming Tidal Power
     from Bay of Fundy Using Compressed Air Energy
     Storage; prepared by Canadian Atlantic Power
     Group Limited for Energy, Mines & Resources
     Canada; Ottawa, Canada.

13.  Anon.; 1981; "Tidal Power from the Severn Estuary",
     (2 vols.); Energy Paper No. 46; Her Majesty's
     Stationery Office; London, England.

14.  Gordon, D. C. Jr., and M. J. Dadswell; 1984; Update
     on the marine environmental consequences of tidal
     power development in the upper reaches of the
     Bay of Fundy; Can. Tech. Rep. Fish. Aquat. Sci.
     No. 1256: vii+686p.

15.  Amos, C. L.; 1988; Personal communication.

16.  Anon.; 1990; LISP project report in preparation,
     Centre for Estuarine Research; Acadia
     University, Wolfville, N. S., Canada.

17.  Solomon, D. J.; 1988; Fish Passage through tidal energy
     barrages, ETSU TID 4056; Energy Technology
     Support Unit, Harwell, Oxfordshire, England.

18.  Baker, G. C., and G. R. Daborn; 1989; Fishery-
     related impacts of the Annapolis tidal generating
     station; Collections Environnement et Géologie
     Vol. 9, pp. 539-554; Canadian Society of
     Environmental Biologists.

19.  Dadswell, M. J., R. A. Rulifson and G. R. Daborn;
     1986; potential impact of large-scale tidal power
     developments in the Upper Bay of Fundy on
     fishery resources of the Northwest Atlantic;
     Fisheries, Vol. II, No. 4.

20.  Ruggles, C. P., et al.; 1990; A critical examina-
     tion of turbine passage fish mortality estimates;
     Canadian Electrical Association research report
     801G658; 57 pp., Toronto, Canada.

21.  Taylor, J. M., B. G. Keefer, and J. Lang; 1985; Ultra-
     low head tidal power and site assessment.

22.  Anon.;          1984; An ultra low head approach to
     tidal energy; H. A. Simons International Ltd.;
     Vancouver, Canada.

23.   Cheng, X.; 1986; Tidal Power in China; Ministry of
      Water Resources & Electric Power; Beijing, China.

24.   Baker, G. C.; 1987; The Competitive Status of Fundy
      Tidal Power; ECE Symposium on the Status and
      Prospects of New and Renewable Sources of Energy;
      Sophia Antipolis, France.

25.   Jones, B. I., et al.; 1992; The Mersey Barrage - Civil
      Engineering Aspects; Tidal Power Conference
      Proceedings, Paper No. 3; Institution of Civil
      Engineers, London, England.

26.   Carnie, C. G., I. D. Jones, I. Hounam, G. Riva, and
      J. Twidell; 1986; Tidal energy potential around
      the coast of Scotland using the Salford Transverse
      Oscillator.  In Energy for Rural and Island
      Communities IV, edited by John Twidell, Ian
      Hounam, and Chris Lewis, pp. 151-158; Oxford,
      United Kingdom: Pergamon Press.

27.   Gorlov, Alexander M.; 1985; High Volume Tidal or Current
      Flow Harnessing System; Washington, DC:  United
      States Patent Office, No. 4,464,080.

28.   White, P. R. S., L. J. Duckers, F. P. Lockett,
      B. W. Loughridge, A. M. Peatfield, and M. J.
      West; 1986; A low head hydro scheme suitable
      for small tidal and river applications.  In
      Energy for Rural and Island Communities IV,
      edited by John Twidel, Ian Hounam, and Chris
      Lewis, pp. 171-177; Oxford, United Kingdom:
      Pergamon Press.

29.   Bellamy, Norman W.; 1989; Low-head hydroelectric power
      using pneumatic conversion; IEE Power Engineering
      Journal, Vol. 3, No. 3 (May), pp. 109-113.

30.   Vauthier, Philippe; 1988; The Underwater Electric
      Kite East River deployment.  In Oceans '88
      Proceedings, pp. 1029 (Vol. 3); Washington,
      DC:  Marine Technology Society.

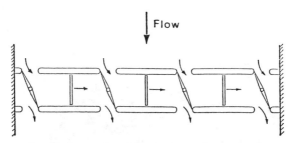

1. Hydrostatic Head Forces
   Paddles To Right

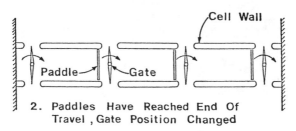

2. Paddles Have Reached End Of
   Travel , Gate Position Changed

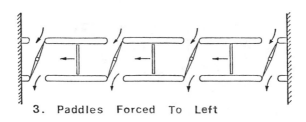

3. Paddles Forced To Left

Figure 1. Salford Transverse Oscillator.

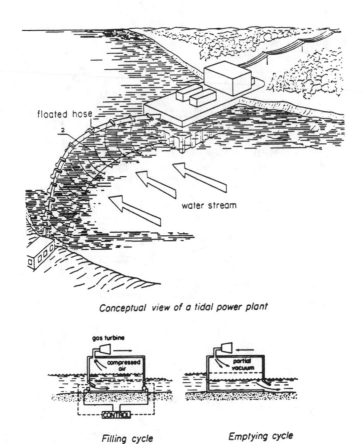

Conceptual view of a tidal power plant

Filling cycle                    Emptying cycle

Figure 2. Pneumatic device developed at Northeastern University.

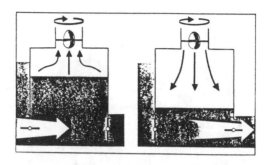

OSCILLATING WATER COLUMN DEVICE

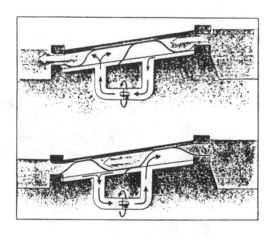

FLEXIBLE BAG DEVICE

Figure 3. Pneumatic devices developed at Coventry Polytechnic.

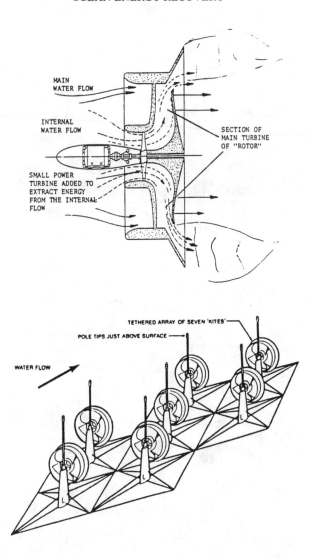

MAIN
WATER FLOW

INTERNAL
WATER FLOW

SECTION OF
MAIN TURBINE
OF "ROTOR"

SMALL POWER
TURBINE ADDED TO
EXTRACT ENERGY
FROM THE INTERNAL
FLOW

TETHERED ARRAY OF SEVEN 'KITES'

POLE TIPS JUST ABOVE SURFACE

WATER FLOW

Figure 4. UEK Corporation's horizontal-axis submerged current turbine.

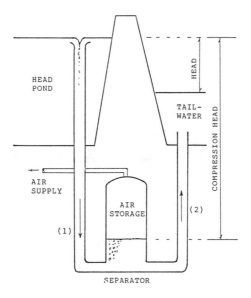

**Figure 5. Hydraulic Air Compressor.**

In the illustration, the hydraulic
head creates a flow through the
intake pipe (1) and the discharge
pipe (2). Air bubbles are entrained
in the intake, removed from the
water column by the separator and
stored at a pressure equal to the
compression head.

# Chapter 5: Closed Cycle Ocean Thermal Energy Conversion.

Dr. F. A. Johnson
Technical Director, GEC-Marconi Research Centre,
Great Baddow, Chelmsford, Essex, CM2 8HN,
ENGLAND.

Table of Contents

# 1 Introduction

The sun's short wavelength radiation falling on the earth is about $173 \times 10^{12}$ kW. About 30% of this is immediately reflected back into space as short wavelength radiation but the remainder - $121 \times 10^{12}$ kW - is absorbed as heat by the oceans, the land and the air from whence it is ultimately re-radiated as long wavelength radiation. Sunlight is very rapidly absorbed in the surface layer of the oceans and this, coupled with convection, creates a huge reservoir of warm surface water up to 100 meters deep. As would be expected the warmest surface water is to be found in the tropics. However, the warm surface water cools by evaporation, especially in the polar regions, and the resulting colder, denser water sinks. This process drives a large, global scale, circulation which keeps the vast mass of deep ocean water at much lower temperatures than the surface. The period of this circulation is estimated to be between 300 and 1,000 years and the heat extraction rate associated with it is estimated to be about $50 - 150 \times 10^9$ kW (1). This value provides a more realistic estimate of the rate of renewal of this ocean thermal energy resource than does the much larger value for the heat input from the sun.

The possibility of utilising the temperature difference between warm surface and cold deep ocean water to drive a heat engine, to produce electric power, was first recognized in 1881 by d'Arsonval in France(2). However, it was also recognized, as a consequence of the second law of thermodynamics, that with the relatively small temperature differences involved it would be necessary to process very large amounts of heat energy to extract any useful work. Thus the challenge is - can one utilise these very large resources of warm and cold sea-water to generate electrical power? It was not until 1929 that the first practical attempt to generate power in this way was made by Claude (3). He utilised an "open cycle" technique - which is described elsewhere - to produce about 22 kW but this was less than the power needed to operate his experimental plant. Thus Claude demonstrated that although he could generate electric power in this way but he could not succeed in generating a surplus.

Interest in the possibilities for exploiting this resource, Ocean Thermal Energy Conversion (OTEC), were renewed in the early 1970's as a result of the abrupt increase in oil prices that occurred at that time. A number of major studies were carried out in the United States of America for the Department of Energy notably by Lockheed (4), the TRW Corporation (5), Westinghouse (6) and the Rand Corporation (7). These studies considered power plants with outputs ranging from 25 to 100 MW. Studies were also carried out in Japan particularly with the establishment of their Sunshine Project in 1974. Japanese studies were directed to plants in the 1 MW to 100 MW range. In Europe studies were initially undertaken by a group of companies from eight nations, called Eurocean. They produced a report on the technical and economic aspects of OTEC in 1977 (8). In many cases this initial work led to the promotion of national programmes. Studies have been conducted by IFREMER, in France, commencing in 1978 and by 1982 they were promoting the design of a 5 MW plant for Tahiti (9). In the United Kingdom studies began in 1981 on a 10 MW floating plant and these studies were led by a company "Ocean Thermal Energy Conversion Systems Ltd." (10).

These studies all concluded that a useful surplus or net electric power could be produced, using the closed cycle process, and that all the engineering techniques needed were within the "state of the art". Considerable emphasis was placed on estimates of the economics of electric power produced in this way. The general conclusion was that power plants of 10 or more megawatts would need to be made and operated in the tropics if the power produced was to be competitive with that produced by diesel or oil fired plant. The unit capital costs of OTEC systems were higher but the unit production costs were expected to become lower than those of fossil fuelled plant especially if the cost of fuel oil continued to increase.

The first proof of the concept of a closed cycle OTEC system was done by a small consortium in Hawaii in 1979 (11). The consortium was formed by Lockheed Missile & Space Co., the State of Hawaii and the Dillingham Co. The plant, called Mini-OTEC, was assembled on a barge and produced an output power of 50 kW and a surplus or net output power of 10 - 15 kW. The tests were run for a period of 3 months and demonstrated that a closed cycle OTEC system could indeed generate gross electric power.

The next experimental work aimed at closed cycle OTEC was done by TRW Inc. and Global Marine Inc. for the U.S. Department of Energy. This project was called OTEC-1 (12,13) and was aimed at testing heat exchangers of 40 MW rating for subsequent use in an OTEC plant. The heat exchangers were shell and tube type using titanium tubes. The tests were successful but the project was cancelled after 4 months as a result of policy changes in the DOE.

In 1980 a pilot plant was built by the Tokyo Electric Power Co., the Tokyo Electric Power Services Co. & Toshiba and Shimizu Construction Inc., in the Republic of Nauru (14), a small island in the Pacific on the equator. This plant was land based and designed for a an output power of 100 kW and a surplus or net power of 11 kW but actually produced an output power of 120 kW and surplus or net power of 31.5 kW. This plant ran for a period of nearly 12 months during which time it delivered power into a local electricity power grid for a period of about two months. The tests were discontinued after a typhoon damaged the cold water pipeline.

Another, but rather different, pilot plant was built in Tokunoshima by the Kyshu Electric Power Co., the Mitsubishi Heavy Industries, Kobe Steel, Tokyo Kyuei Co., & the Tobishima Construction Inc. (15). This plant was designed to produce 50 kW power using warm water from a diesel electric set rather than surface water. The warm water was at a temperature of 40.5°C. The plant produced some 28 to 38 kW but the warm water flow rate was less than anticipated. This type of plant is really a "bottom cycle" system since one is exploiting waste heat from a conventional power plant to produce additional electric power.

It will be seen from the foregoing that, although the initial idea is very old, the main activity has come about recently as a response to the sharp rise in oil prices in the early 1970's. However, part of the response to this sudden change was a crash programme to produce, as rapidly as possible, designs for large scale OTEC plants using, of necessity, existing technology. This in turn led to the conclusions that closed cycle OTEC itself was "state of the art" technology and hence needed little further development to become a commercial reality. Further, the studies indicated that the technology for a closed cycle OTEC system was a high capital cost technology which would not become competitive with oil unless oil prices continued to rise. As a consequence, with the fall in oil prices in the early 1980's, most of the interest in closed cycle OTEC disappeared.

The main reasons for the high unit capital costs of closed cycle OTEC can be summarised as follows:

i)     the very low Carnot efficiencies of the closed cycle OTEC process results in the need for very large and hence expensive heat exchangers;

ii)    the high cost of the sea-water systems;

iii)   the substantial reduction in the surplus or net electric power due to the power consumption of the sea-water pumps.

This is a very daunting list of disadvantages which is being further compounded by the rapid improvement in the efficiency of modern, combined cycle, power generation plant which uses natural gas as the primary source of heat. These new plants have unit capital costs of about $700 per kilowatt and conversion efficiencies of over 50%. The question thus arises can OTEC have an economic future? I believe it can but not, at least for a considerable time, in the context of pure electricity production.

In retrospect, one can see that although the work on the pilot plants demonstrated the proof of the concept, the design studies indicated that if an OTEC plant was to produce electricity in a cost competitive way the net output would need to be increased from the 30 kW of the Nauru demonstration to 10 MW or more, the availability increased to about 80% or better and the plant operated for about 25 years or more. Further, the unit capital costs would need to be reduced substantially. This would involve a very large extrapolation from proven technology and consequently there would be a high risk of meeting unforeseen problems which could prove to be very costly.

From a technical point of view, what is needed is a more modest scaling up of the previous American & Japanese pilot plants. A closed cycle land based plant with a gross output of 150 to 1,000 kW connected to a distribution grid and operating for a minimum of 5 years with a high availability would provide a very thorough test of the whole OTEC concept and, in particular, the materials and fabrication techniques. Such a demonstration would provide a sound basis on which to plan and design larger OTEC installations.

However, as the design studies have shown, a plant of this type would be uneconomic. This highlights the difficulties for closed cycle OTEC namely:

i)     the economics of power generation suggests that a very large technical leap forward is needed;

ii)    engineering prudence suggests that an incremental approach is the soundest way to proceed;

iii)   the problems of financing an incremental approach are complicated by the fact that small electric power plants are uneconomic;

iv)    the problems of financing a large leap forward are bound to be considerable in view of the risks of unforeseen problems.

This article will first discuss the basic technology of closed cycle OTEC and then review the practical engineering developments that have been taking place during the 1980's which have steadily increased the confidence in and reduced the costs of closed cycle OTEC technology. It will then outline the very important work that has taken place at the Natural Energy Laboratory of Hawaii which has been concerned with the development of a variety of co-products, especially mariculture, based on the exploitation of deep ocean water. This mariculture work has, in turn, opened a niche market for small closed cycle OTEC plants to provide the electric power for the sea-water pumps and other operations as an alternative to buying electric power from a local grid. Finally this article will review some of the environmental issues for closed cycle OTEC.

## 2 Thermodynamics.

### 2.1 Basic Concepts.

We should remember that if one could devise a reversible heat engine with a working fluid which took in heat at the temperature of the surface sea-water and discharged heat at the temperature of the cold deep sea-water the power output would be completely defined by the first and second laws of thermodynamics. The second law of thermodynamics states that, for a reversible heat engine, the entropy taken in at the input temperature is equal to the entropy discharged at the output temperature and that the power produced, by each unit mass of working fluid, is its entropy per unit mass times the temperature difference between the input and the output. This is summarised by the well known Carnot efficiency equation which defines the theoretical maximum fraction of the input heat that can be converted into useful work by a reversible heat engine.

A practical closed cycle heat engine for OTEC uses a working fluid that is chosen for its more convenient thermodynamic properties when compared with sea-water. As a consequence the closed cycle OTEC system needs two heat exchangers, one the evaporator, where heat is transferred from the warm sea-water to the working fluid and the other, the condenser, where heat is transferred from the working fluid to the cold sea-water. In all closed cycle OTEC demonstrations a Rankine cycle was used as the nearest practical approximation to an ideal heat engine. The details of a Rankine cycle are discussed more fully in section 2.3 and an outline of the system is shown in figure 1.

Finally a word about OTEC terminology. A thermodynamic cycle involves both positive and negative work, the difference being the net work. The ratio of the net to positive work is the work ratio which, for a Rankine cycle, is large. If a closed cycle OTEC plant were to be used as a primary source of electric power, the convention is that the fuel would be free but the power needed to drive the sea-water pumps is taken as a further contribution to the negative work of the cycle so that the net power of the cycle is reduced by this pumping power. The net power of the Rankine cycle itself (i.e. excluding the pumping power) is then called the *gross* power to distinguish it from the *net* power which includes the negative work of the sea-water pumps. However, if an OTEC plant is installed to provide the power for operating the sea-water systems of a large mariculture operation and, in addition, to provide a useful surplus, the *net* power should not be reduced by the sea-water pumping power - the OTEC plant is simply an alternative to buying electricity from

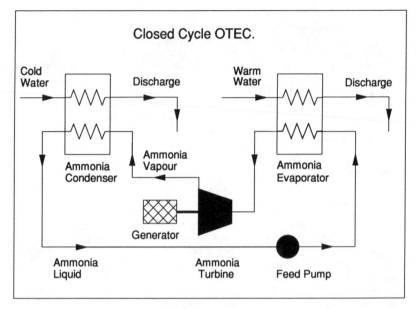

## Closed Cycle OTEC.

Figure 1.

a local grid or buying a diesel electric generating set. To simplify the discussions, when dealing with these two very different applications, I shall use the term *gross* power for the output of an ideal or Rankine cycle and *net* or *surplus* power for the difference between this and the sea-water pumping power. Thus in the case of an OTEC plant used for generating electricity alone, the *net* or *surplus* power is the measure of the output, while in the case of the OTEC plant used as an adjunct to a larger mariculture operation, the *gross* power is the measure of the output.

### 2.2 The Theoretical Gross Power Output.

Sections 9.1, 9.2 and 9.3 in the appendix discuss the physics of heat transfer to and from the working fluid and examine the maximum power that could be produced if the Rankine cycle, itself, worked as a perfectly reversible heat engine. In the analysis the heat exchanger performance of the evaporator is defined by the dimensionless parameter $X_e$. This is the ratio of the thermal conductance times the area of the heat exchanger, $(U_e A_e)$ to the specific heat times the mass flow of the sea-water, $(C_{pw} \dot{M}_{ew})$ - see equations 9.2.3 and 9.1.3. This in turn defines a dimensionless parameter $Y_e$ for the evaporator as:

$$Y_e = 1 / [1 - \exp(-X_e)] \qquad 2.2.1$$

Similarly for the condenser:

$$Y_c = 1 / [1 - \exp(-X_c)] \qquad 2.2.2$$

It is then shown - equation 9.3.5 - that the condition for the maximum gross power output from such a reversible heat engine occurs when the temperature reduction of the warm sea-water passing through the evaporator, $\Delta T_e$, is given by:

$$\Delta T_e \left( Y_e + Y_c \, G_e / G_c \right) = \tfrac{1}{2} \Delta T_i \left[ 1 + \tfrac{1}{4} C_o \left( 1 + \delta \right) \right] \qquad 2.2.3$$

where $\Delta T_i$ is the initial temperature difference between the warm and cold sea-water, $C_o$ is the theoretical Carnot efficiency defined by the equation:

$$C_o = \Delta T_i / T_{ei} \qquad 2.2.4$$

where $T_{ei}$ is the initial temperature of the warm sea-water and where:

$$\delta = \tfrac{1}{2} C_o \left[ 1 + \tfrac{5}{8} C_o + \tfrac{5.7}{8.10} C_o^2 + .. \right] \qquad 2.2.5$$

It is also shown that this result implies that just under half the initial temperature difference between the warm and cold sea-water appears across the heat engine and the remainder is used to drive heat through the heat exchangers - equation 9.3.10. Thus the overall conversion efficiency is very nearly halved by these irreversible heat transfer processes. Finally it is shown that the maximum gross power, $P_m$, is given by the equation:

$$P_m = \frac{\Delta T_i^2 \left( 1 + \delta \right)}{4 \left( Y_e / G_e + Y_c / G_c \right) T_{ei}} \qquad 2.2.6$$

It will be seen that the maximum gross power is proportional to the square of the initial temperature difference and to the mass flow rates of the sea-water. The dependence of the maximum gross power on the evaporator and condenser is through the parameters $Y_e$ and $Y_c$. As the values of $Y_e$ or $Y_c$ increase the fraction of the heat extracted from the sea decreases and, with it, the gross power.

In table 4 (shown in section 9.3) an example is given of the calculation of the maximum gross power output for a closed cycle OTEC plant with $X_e = X_c = 1.0$ and with sea-water flows of one cubic meter per second (15,856 US gals/min or 13,198 Imp gals/min). It will be seen that the maximum gross power is 543 kW. The effect of increasing the performance of the heat exchangers, while keeping the other parameters of table 4 constant, can be seen in table 1 where the maximum gross power output is shown for a series of values of $X_e$ which, in each case, is taken to be the same as $X_c$.

It will be seen, from table 1, that as the value of $X_e$ is steadily increases the value of $Y_e$ decreases asymptotically to unity and the maximum gross power increases asymptotically to 859 kW. This highlights a major problem in the design

The Variation in Gross Power Output with Heat Exchanger Performance
for a 1 cubic metre per second Flow Rates.

| $X_e$ | $Y_e$ | $P_m$ |
|-------|-------|-------|
| 0.5 | 2.5415 | 338 kW |
| 1.0 | 1.5820 | 543 kW |
| 1.5 | 1.2872 | 667 kW |
| 2.0 | 1.1565 | 743 kW |
| 2.5 | 1.0894 | 789 kW |
| 3.0 | 1.0524 | 816 kW |
| 3.5 | 1.0311 | 833 kW |
| 4.0 | 1.0187 | 843 kW |
| $\infty$ | 1.0000 | 859 kW |

Table 1.

of a closed cycle OTEC system namely that the cost of the heat exchangers is largely
proportional to $X$ but that the gross power output from the plant is not. It is of interest
to note that the value of $Y_e$ used in the Japanese pilot plant in Nauru (14) was 2.0.

## 2.3 The Rankine Cycle.

### 2.3.1 The Idealised Case.

As indicated above the form of heat engine cycle most suitable for closed
cycle OTEC system is the Rankine cycle. The details of a Rankine cycle are
discussed more fully in section 9.4 but an outline of the system is shown in figure 1.
The working fluid is heated in the evaporator where it boils and generates a high
pressure vapour. In its simplest form the vapour is saturated so that its pressure is
a function of its temperature only. This high pressure vapour then expands
isentropically through the turbine so producing mechanical power and the low
pressure exhaust vapour is condensed back to liquid form in the condenser. The
vapour in the condenser is also saturated so its pressure is a function of its
temperature only. The liquid working fluid is then isentropically compressed by
the feed pump into the evaporator for re-heating and re-cycling. As the specific
volume of the liquid form of the working fluid is very much less than that of the
saturated vapour form, the power used by the feed pump is correspondingly very
much less than the power produced during the expansion of the vapour in the turbine.
In fact this is the characteristic feature of the Rankine cycle. It utilises a phase
change between the expansion and compression parts of the cycle to reduce the
negative work needed during the compression to a small fraction of the positive
work produced by the expansion so that the work ratio, the ratio of the net work
(or in OTEC terminology the gross work) to the positive work, is very high. If a
gas is used in a heat engine the negative work used in the compression phase is a
larger fraction of the positive work produced during the expansion phase so that
the work ratio is much smaller.

It is important to note that in the evaporator the heat is supplied at constant pressure - the saturated vapour pressure at the output temperature from the evaporator. Similarly in the condenser the heat is extracted at constant pressure - the saturated vapour pressure at the input temperature to the condenser. The expansion of the saturated vapour through the turbine is, ideally, isentropic and similarly the compression of the liquid by the feed pump is also, ideally, isentropic. However, the heating of the working fluid leaving the condenser and entering the evaporator is an additional source of irreversible loss. This results in the conversion efficiency falling from 3.74% for the ideal heat engine to 3.63% and the gross power falling from 543 kW to 533 kW. The details of the performance of the Rankine cycle are given in section 9.4.1. Table 5 gives the calculation for a Rankine cycle using the same input data as in table 4 but, in this case, using ammonia as the working fluid.

## 2.3.2 The Actual Case.

The arguments in the previous section were all based on the assumption that the expansion through the turbine was isentropic - that is it was reversible - and was thus producing the maximum conversion of heat energy to work. Actual expansion turbines do not achieve this ideal performance and the actual work produced is less than the theoretical amount. This is discussed, in more detail, in section 9.4.2. Table 6 lists the modifications that result if an 85% isentropic expansion efficiency is assumed for the turbine. It will be seen that this reduces the conversion efficiency from 3.63% for the ideal Rankine cycle to 3.07% and the gross power from 533 kW to 451 kW.

## 2.3.3 Practical Considerations.

In the discussion above of the simple Rankine cycle, using dry saturated vapour as a working fluid, we have taken account of the major losses of performance - the irreversible transfer of heat, the isentropic efficiency of the expansion turbine and the irreversible losses associated with heating the working fluid as it enters the evaporator. In addition the gross power is further reduced by friction losses as the working fluid flows around the system and losses in the power chain between the expansion turbine and the feed pump. However, these reductions in the gross power can be kept to relatively low values.

Another reduction in the gross power is the sea-water pumping losses that arise from the back pressure head in the heat exchangers themselves. This is an important design feature since the heat transfer rate is markedly affected by the turbulence of the sea-water flow past the heat conducting surfaces. In the nearly laminar regime both the back pressure head and the heat transfer rate are low but as the flow becomes turbulent both of these quantities increase markedly.

Yet another reduction in the gross power arises from the loss is in the conversion from mechanical power to electrical power and vice versa. Typically the efficiency of the conversion from mechanical to electric power is 90% to 95% but the reverse conversion is lower, especially for small amounts of mechanical power. The reasons for these losses are outside the scope of this article but they do, of course, effect the overall system performance.

A disadvantage of working with saturated vapour is that during the evaporation process a mist of small drops of liquid can become entrained in the vapour flow and this can, in time, cause erosion of the turbine blades. The usual way to minimise this problem is to include a simple mist separator in the vapour stream before it enters the turbine. This is a low loss filter that traps a significant portion of the entrained droplets into a liquid pool which can then be fed back into the evaporator input. Another problem is that the saturated vapour cools during the expansion through the turbine, especially during the initial expansion through the inlet nozzles. This can generate a stream of high velocity liquid droplets that impinge on the rotor blades and gradually cause erosion. This is a less serious problem in an OTEC plant since the dryness fraction, a measure of this condensation process, is very high.

In seeking to improve the efficiency of an OTEC plant we may do well to look at the methods frequently used to improve the performance of the Rankine cycle in large steam turbine plants: namely superheating, reheating and preheating. Of these three, reheating is not applicable to OTEC, since, with the low pressure drops involved, only a single stage turbine is used, but superheating and preheating can have roles to play.

Superheating is a technique for directly heating the saturated vapour from the evaporator from the highest temperature source available. This additional heating occurs at constant pressure and it increases the enthalpy of the, now superheated, vapour. This not only increases the energy conversion by the turbine but also reduces the problem of condensation during expansion from the inlet nozzles since the temperature has firstly to be reduced to the saturated vapour pressure before any condensation can begin. In addition entrained liquid droplets also become converted to vapour. Although in an OTEC plant there is less scope for superheating its obvious advantages can be worth exploiting where possible. For example, in the case illustrated in table 4 the evaporator temperature is $5°K$ below that of the incoming sea-water so that it should be possible to provide some 3 to $4°K$ of superheat in order to reduce erosion problems in the turbine.

If, in the above example, a small superheater, with an $X$ value of 0.01, was placed in the incoming sea-water before it entered the main evaporator, the ammonia vapour temperature would be increased by $4.00°K$, by the absorption of 134 kW of heat. This increases the dryness factor from 97.0% for the ideal Rankine cycle to 97.9% for the superheated cycle but, in spite of a significant increase in the temperature difference across the turbine, there is virtually no increase in the conversion efficiency. Superheating should be seen primarily in the context of removing moisture droplets from the vapour and reducing erosion problems in the expansion turbine. However, if there is an external source of additional heat, for example solar energy or a salt pond, its use to superheat the ammonia vapour could be considerably more effective.

Preheating is a process also widely used in large steam turbine plant where some of the vapour from intermediate stages in the expansion process is extracted and used to progressively heat the feedwater (working fluid), often in several stages, before it enters the boiler (evaporator). By careful design this preheating process can ensure that the feedwater (working fluid) leaving each pre-heater stage is nearly at its saturated liquid temperature. This more controlled heating of the feedwater (working fluid) makes a useful gain in the overall conversion efficiency of the system by reducing the irreversible losses.

Preheating can be readily applied to an OTEC plant since the water leaving the evaporator is at a significantly higher temperature than that of the working fluid leaving the condenser. In the example given in table 4 the sea-water leaving the evaporator is some 12°K warmer than the liquid ammonia leaving the condenser. If a heat exchanger, with an $X$ value of 0.02, is placed in the water leaving the evaporator, the temperature of the ammonia would be increased by 9.91°K by the absorption of 561 kW of heat. The effect of this preheating is to raise the cycle efficiency from 3.07% to 3.11% - a useful increase. Thus preheating does provide gains in both the heat absorbed and in the conversion efficiency of the process. It is worth noting that Mini-OTEC (11) used a separate preheater. Clearly the combined effect of both superheating and preheating will increase the dryness factor and the conversion efficiency.

## 2.4 Working Fluids

Claude (3), in fact, used the sea-water itself as a working fluid and so avoided the need for heat exchangers. The reason for using a different working fluid arises from the thermodynamic properties of water vapour at the temperatures involved. Table 2 shows the saturated vapour pressure of steam at temperatures of 26°C and 6°C - figures not un-typical of ocean surface and deep water. It will be seen that the pressures are very low and the specific volumes are very large.

Thermodynamic Properties of Steam.

| Temperature | Pressure | Specific Volume |
|---|---|---|
| 26 °C | 3.36 kPa | 9,033 l/kg |
| 6 °C | 0.94 kPa | 30,400 l/kg |

Table 2.

This presents a major technical problem in that such an "open cycle" system has to run at a near vacuum pressure and the heat engine has to be designed to work with very large volumes of low pressure steam. By contrast ammonia has the thermodynamic properties for the same two temperatures shown in table 3.

Thermodynamic Properties of Ammonia Vapour.

| Temperature | Pressure | Specific Volume |
|---|---|---|
| 26 °C | 1,034 kPa | 124.5 l/kg |
| 6 °C | 535 kPa | 235.1 l/kg |

Table 3.

It will be seen that, by contrast with steam, the pressures of ammonia vapour are much higher and the specific volumes are much lower. This greatly simplifies the design of a suitable heat engine for use in a "closed cycle" system. In fact the reason for using a closed cycle system is that the advantages of using such a relatively high pressure heat engine can outweigh the disadvantages and inefficiencies associated with the use of heat exchangers. It will be recalled that, although the first gross power produced by OTEC was made with an "open cycle" system - Claude (3) - this attempt failed to produce surplus or net power. The first practical demonstration of the production of surplus or net power - Mini-OTEC (11) - was made using a closed cycle system with ammonia as the working fluid.

The basic requirement for a good working fluid in an OTEC plant is that it has a high vapour pressure at the upper temperature as this ensures a good pressure drop across the expansion turbine and hence makes it small and efficient. Ammonia, propane and a number of refrigerant liquids such as R-22 fulfil this requirement. The pressure ratios between the upper and lower temperature is less dependent on the choice of working fluid (Trouton's Rule) so the final choice must be made from other considerations such as the detailed thermodynamic properties, performance aspects in the particular system being considered, compatibility with materials, safety aspects or environmental considerations.

The Japanese considered no fewer than twelve possible working fluids for a 25 MW design as part of their "Sunshine Project" and, although ammonia, propane and R-22 were found to be almost equally suitable from a system point of view, ammonia stood out as the best choice when the size of the heat exchangers and piping costs were taken into account (16). Our own work came to precisely the same conclusion for the same reasons (17). However, in spite of this assessment, the Nauru project team chose R-22 for their 100 kW plant was R-22 (14), because of worries over the flammability and toxicity of ammonia. However, the other Japanese group used ammonia in their 50 kW plant at Tokunoshima (15) - presumably the worries over the use of ammonia were found to be of less importance than previously assumed.

## 2.5 Expansion Turbines.

The pressure changes developed by a closed cycle OTEC plant is, by most power engineering standards, low and hence the turbine must be able to operate with relatively high mass flows of the working fluid. For this reason it is unnecessary to consider more than one expansion stage. Further, as the gross power output varies as the square of the initial temperature difference, the output of an OTEC plant can vary by more than ± 20% over the course of a year so that the turbine needs to be designed to have a good isentropic expansion efficiency over a considerable range of pressure variations.

This requirement is best met by the use of radial inflow expansion turbines with variable inlet nozzles. This type of expansion turbine can have an isentropic expansion efficiency of 85%. The use of variable inlet nozzles allows them to maintain high expansion efficiencies over a very wide range of mass flow rates - in excess of 80% for variations in the mass flow of ± 40%. Radial inflow turbines also have much sturdier blades than those used in axial turbines so they have a significantly greater resistance to moisture erosion problems.

The Rotoflow Corporation have developed a range of turbines of this type for power outputs of 75 to 10,000 kW. These turbines have been developed for the process industries and they can handle a very wide range of working fluids. The sealing arrangements, in particular, are very good and these turbines can be used safely even with hydrogen gas. Technically these radial inflow turbines are well suited for use in a closed cycle OTEC plant. However, they are designed for the much larger pressure differences that are met with in the process industries so that they are over-designed for OTEC applications. There are, of course, considerable economies of scale in the unit price of such turbines - the unit costs ranging from about \$2,000/kW at the low power end to \$300/kW at the high power end of the range.

## 2.6 Hydraulics

The process of pumping the large volumes of sea-water either directly for an OTEC plant or for a large mariculture operation uses a large amount of electric power. This reduces the surplus or net power that can be produced by an OTEC plant significantly. In addition the process of pumping the deep cold water through pipelines leading to the surface through warmer surface water increases the temperature of the cold water which reduces the thermodynamic efficiency of the OTEC plant. Both of these issues are discussed in section 9.5.

The gross power output of an OTEC plant is proportional to the mass flow rates of the sea-water and the pumping power required is proportional to the mass flow rate and the hydraulic back pressure head. It is shown that the hydraulic suction head due to flow resistance in the pipeline is approximately given by equation 9.5.1:

$$H_s = A_o \, (L \, / \, \rho) \, (Q^2 \, / \, d^5) \qquad\qquad 2.6.1$$

where $H_s$ is the suction head, $A_o$ is a constant for the particular type of pipe, $(L \, / \, \rho)$ is the length of the pipe divided by the density of the water, $Q$ is the water volume flow rate and $d$ is the diameter of the pipe. Thus to minimise this suction head one should use large diameter pipes. However, such pipes are more expensive than small diameter pipes so there has to be a compromise between the capital cost and pumping power.

It is also shown, in section 9.5, that the temperature rise associated with the warming of the cold water by the surroundings on its way to the surface depends on the ratio of the pipe diameter to the mass flow rate. Thus the heating of the cold water diminishes with increasing mass flow rates but, for a given mass flow rate, the heating effect increases with the diameter of the pipe.

## 3 Recent Developments

In section 1 we reviewed the development of closed cycle OTEC up to the early 1980's and in section 2 we discussed the basic thermodynamics and the design problems that arise in a closed cycle OTEC power conversion plant using the Rankine cycle. We also touched briefly on some of the design aspects of the sea-water systems. In the remaining sections we shall consider the sea-water systems, the economic aspects and the current perception of closed cycle OTEC. In each of these areas ideas have been changing and developing. Finally I will give some personal views of the possible course of future developments.

## 3.1 Sea-Water Pipes

As discussed in section 2.2 the gross power output from a closed cycle OTEC plant is proportional to the flow rates of the sea-water and the square of the initial temperature difference. Thus decreasing the initial temperature difference from 22°K to 20°K, in the previous examples, would decrease the gross power by 17.4%. In addition, the seasonal variation in the output becomes less the greater the mean temperature difference. Now the temperature of the surface sea-water depends on the geographical location and the time of the season and is an input parameter for a particular plant. However, one should try to obtain the coldest sea-water available at the site by making the cold water pipe reach to sufficiently great depths. Sea-water, unlike fresh water, reaches its maximum density almost at its freezing point, -1.9°C, so the temperature of ocean water decreases with depth but the rate of decrease becomes smaller the greater the depth. At depths of about 1,000 meters the temperature is about 4°C. The choice of depth for the cold water pipe is a compromice between increasing costs and increasing efficiency.

### 3.1.1 Floating Systems.

The earliest demonstration of closed cycle OTEC, Mini-OTEC (11), was made from a barge. This was in Hawaii in 1979. The warm and cold sea-water pipes were suspended from the barge as was a discharge pipe. This meant that the pipes were the shortest possible length to reach the required depths. This makes the minimum demands on the sea-water pumps and, in addition, it makes relatively simple demands on the pipe material and its fabrication.

The OTEC-1 (12),(13) experiments also used a floating platform. These experiments too were carried out in Hawaii later in 1980. In this case the warm water was brought in directly from the ship's bottom at a flow rate of 5.2 m³ per second. The cold water pipe was made up from a cluster of three polyethylene pipes, 686 meters long, suspended from a gimbal mount in a moon pool. The individual pipes had a diameter of 1.22 meters and the total flow rate of the cold water was 4.3 m³ per second. The gimbal mount displaced 20 tonnes in the water and could support 195 tonnes in the vertical and could withstand thrusts of 36 tonnes in the horizontal. A polyethylene coated wire rope passed down through the cluster and on to a weight, some 90 meters lower, to offset the small positive buoyancy of the polyethylene. Finally, there was a single pipe about 66 meters long, suspended from a gimbal, through which the mixed warm and cold sea-water was discharged. At this depth the discharge was neutrally buoyant. Polyethylene has a low thermal conductance and can be heat welded relatively easily to make long continuous pipes. It is also highly resistant to sea-water. It is an elastoplastic material so that it has a complex behaviour for strains beyond the normal elasticity range. As a consequence of this one needs to take considerable care in devising a deployment strategy, for positioning these very long pipes, that does not permanently modify the elastic properties of the pipe. This was quite a difficult manoeuvre involving four steps:

1)      the pipes were first towed out to the barge in a horizontal position, by a tug;

2)      a line and weight were attached to the free end of the pipe from the barge;

3)      the pipe was then upended by lowering the line from the barge while the tug maintained a fairly constant tension along the pipe during this process so as to avoid excessive bending;

4)      when in the vertical position alongside the barge it was then keelhauled into position.

The whole process was successful but it illustrates the careful planning needed to accomplish such a task. This pipe was tested for a period of about 5 months.

A number of design studies of large closed cycle OTEC plants have favoured this concept. The advantages include short pipes, small, if any, lifts of the water above sea level, and relatively straightforward deployment schemes. The problem is that such a plant has to include a floating platform, the means for mooring it in place in deep water, possibly means for limiting the lateral movement of the suspended pipe. The lateral movement of such a barge attached to a deep tether can be quite large - 2.3 kilometres in the case of OTEC-1. This lateral movement as well as the depth and distance from shore greatly complicates the problem of delivering the surplus or net electric power to land where it is normally required. These latter problems do not arise if the plant is located on land and the sea-water brought to the plant.

3.1.2 Land Based Systems.

The earliest attempt to install a cold water pipe from the land was made by Claude (3). He made up a 1.6 meter corrugated steel pipe with rubber joints and installed it, after one or two unsuccessful attempts, by pulling it out from the shore with ballasting floats and then allowing the pipe to sink in a somewhat but not entirely controlled fashion. The whole deployment operation took only a few days. However, as a result of his experiences Claude suggested that a floating concept might be better or that one should build a tunnel out from the land!

The next land based installation was the Nauru project in 1981 (14). Here the cold water pipe was a 932 meters long, 0.70 meters diameter polyethylene pipe with a wall thickness of 30 mm increasing to 35 mm near the shore end. This was to minimise the thermal conductance of the pipe in the region of the greatest temperature difference with the surrounding water. The pipe was fabricated by welding together 10 meter sections. Weights were attached at regular intervals, each of which was chosen to give a margin of safety against the current forces likely to be encountered at its eventual depth. The pipe was deployed by suspending it from a hanger wire attached to floats and pulling it out from the shore with a tug. When the pipe was stretched out on the surface a second tug attached a line to the mid point of the hanger wire, the floats were detached and the pipe allowed to slowly submerge. The two tugs controlled the tension in the hanger wire to minimise the bending of the pipe. In position the pipe lay on the sea bed. The deployment was completed in less than three days. This pipe gave good service for nearly a year but was then destroyed by a typhoon.

Both of the systems just described made use of bottom mounted pipes. These should work well provided that the sea bed, where the pipe eventually settles, is relatively smooth so as to provide a good distributed support. However, this is often not the case. Even when modern underwater surveys are available, there is still a major problem in establishing precise coordinates for the survey data and a further problem in deploying the pipe precisely onto the selected route. The experience of some more recently installed bottom mounted pipes (18) has been that in some cases the pipe landed on a large underwater cliff so that very sharp bends were induced which resulted in a relatively short life.

These problems with bottom mounted pipes were overcome by an ingenious design by Van Ryzin (18) of Makai Ocean Engineering, Hawaii. Recognizing that polyethylene is positively buoyant and that it has an excellent strain capability, he designed and installed a 915 meter long and 0.3 meter diameter pipe in which the major section formed a natural inverted catenary thus keeping it well clear of the bottom. At the near-shore end the pipe was attached to a gimbal which formed the transition section between the catenary and a bottom mounted pipe leading to the shore. Two anchors were used, one at the transition section and the other at the intake end. At the time it was thought that the ocean currents would be small and that the design life would be one year. The pipe was installed at Keahole Point in Hawaii in 1981 and has been in operation ever since. This is all the more remarkable as the ocean currents were later found to be much stronger than had been thought. The method of deployment is described below.

Van Ryzin (18) also designed and installed a larger 1.06 meter polyethylene pipe at Keahole Point for a flow rate of 0.84 m$^3$/s. Here the pumps were in a sump 76 meters inshore. For the first 92 meters off-shore the pipe was buried. From this point, at a depth of 21 meters, for the next 640 meters, to a depth of 152 meters, the pipeline was held just off the bottom with concrete anchors clamped to the pipeline. Beyond this point for the next 1,376 meters the pipe formed an inverted catenary reaching to a depth of 640 meters. The catenary was anchored at the top, or transition, end with a 36 tonne anchor and at the bottom with an 18 tonne anchor . The load was taken from an external pipe clamp by a line to the anchors. In the shallow region of the catenary additional buoyancy was added to help with the high currents of 1.3 m/s. These currents decrease to about 0.4 m/s at depths of 305 meters or below. In the strongest currents the centre of the catenary section moves about 152 meters horizontally and 76 meters vertically but this is within the design specification.

The pipe was fabricated at a nearby harbour in two sections. The first section was 1131 meters long and included the near shore section. The near shore section had weights attached and the assembly was floated, air-filed, in the harbour. It was then towed to the site, anchored at the shore end and attached to shore pumps. A tug pulled the pipe into alignment and the transition anchor was attached. Then, by pumping water into the pipeline at the shore end, the pipeline was submerged to the bottom. By controlling the air pressure inside the pipe, the pull from the tug and the flow of water into the pipe, the pipe could be kept in static equilibrium and thus the submergence could be controlled. The transition anchor was also lowered as the pipe submerged. Most of the remaining pipe, 366 meters beyond the transition section, remained floating. The second section, 610 meters long, was towed out from the harbour and the two flanged pipe sections were bolted together on the surface. A pig was pumped through the pipeline to purge it of air. Finally the intake end was attached to the lower anchor which was then lowered to the bottom some 700 meters below the surface. This pipe was installed in July 1987 and continues to give excellent service. The design was conservative and was for a life of 10 years. If the experience of the first, 1981, pipeline is repeated this new pipeline should have a very long life indeed.

The advantage of this technique is that, apart from the near-shore section, one does not need a precise survey of the site nor does one need to be very precise in positioning the pipe during deployment since the major part of the pipeline is well clear of the bottom. Further there are only two anchors to deploy. However, the disadvantages are that these anchors need to be very substantial and the transition region is complex. The following year Makai Ocean Engineering (18) deployed, at a nearby site, another pipeline 1830 meters long and 0.43 meters diameter. In this case a more favourable route had been identified and a pendant design was used. In this the major part of the pipeline was attached, at regular intervals right down to the intake, to a series of anchors by lines of 12 meters length. This kept the pipe well clear of the bottom thus eliminating the need to find a smooth bottom route. To a large extent the pipe follows the bottom contour so the route chosen must not result in any sharp bends in the pipe.

This pipeline was assembled at the site, placed on rollers and launched into the ocean through a ramp at the shore-line. As the pipe was pulled out to sea, pendants, weights and other components were attached near the shore-line. The total launch took about 12 hours and the pipe was laid on the bottom using the same controlled flooding and submerging techniques described above. The submerging took less than 12 hours. To ensure that the pipe landed in the correct position shore ranges were monitored and restraining lines were moored to the bottom at key locations to guide the pipe into the correct position. This pipeline too continues to give excellent service.

With the exception of Claude's (3) corrugated steel pipe, all these pipes were made from polyethylene. From this in might appear that polyethylene is the standard material and that it will only be a matter of time before pipes of any length or diameter will be deployed. This is not the case. If larger diameter pipes were needed there is firstly the difficulty of supply - the largest diameter polyethylene pipe currently made is 1.6 meters - and secondly, with increasing diameter the pipes become more rigid and the techniques described above, for land based plants, become inappropriate. Van Ryzin's view is that the maximum diameter that could be used with polyethylene pipes and the inverted catenary design is, in fact, 1.6 meters (18).

During the 1980's a number of design studies addressed the problem of installing much larger diameter cold water pipelines. These studies are well summarised in a publication produced as a result of a workshop on Bottom Mounted OTEC Sea-water Systems (19). Most of these studies considered materials other than polyethylene. Fibre reinforced plastic, either as a single pipe or as a series of pipes with flexible bellow joints at regular intervals, steel, concrete and aluminium pipes were examined. These larger pipes would require a major effort to deploy - anything up to 6 months. However, one of these studies, by Wenzel of Marine Development Association, produced an entirely new concept - the "soft pipe" - which is briefly described below.

Wenzel argued that if the wall thickness of the pipe could be reduced, the pipes would become more flexible and thus much large diameter pipes could be deployed but the pump suction head would have to be reduced which would reduce the flow rate. However, if the pumps were placed at the intake end, the pipe would be operating under pressure and a thin walled pipe would then be satisfactory. He proposed using a hypalon elastomer material, reinforced with E-glass, of 2.5 to 5 millimetres thickness for the pipe fabric and a pendant type mounting, similar to that described above, with the pipe being retained inside space frame modules attached to the pendant anchors with a spacing of about 15 meters between the pendants. The pendant anchoring system is especially important in the case of a "soft pipe" as it is more susceptible to damage if it came into contact with the sea bed.

### 3.1.3 Warm Water and Biofouling.

Warm water pipes are much simpler to install. Experience at Keahole point, Hawaii, indicates that the intake for the warm water pipe should be in water at least 30 meters deep with the intake some 20 meters below the surface. This is to avoid excessive biofouling and other forms of contamination. Biofouling, the growth of marine organisms in the water system, is primarily a problem with warm water. It can rapidly reduce the thermal conductance of heat exchangers and is not easy to clean off. The best technique is to prevent its occurrence and have a mechanical cleaning technique as a back-up. Small doses of chlorine, 70 parts per billion, for 10 minutes each hour, have been found to prevent biofouling (20)(21)(22). If biofouling is allowed to build up, these same low doses of chlorine can largely cure the problem but, in the case of heat exchangers, mechanical cleaning must also be used if they are to recover their original thermal conductance. Biofouling is much less likely to occur in cold water, but again, low doses of chlorine can prevent the problem. These low levels of chlorine have no effect on fish life.

### 3.1.4 Sea-Water Pumps.

The pumps required for OTEC plants have to generate large flow rates. The factors which define the pressure head they have to work against are discussed briefly in section 2.6. For satisfactory start up the absolute water pressure at the pump inlet must be larger that its design requirement. A simple way to achieve this is to install the pump in some 7 to 10 meters depth of water either in an on-shore sump or off-shore. With very long cold water pipes, starting up can be a problem as a very large mass of water has to be accelerated to the required flow velocity and this can take many minutes. With a "soft pipe" there is the additional need to modify the design of the sea-water pump to make it operate at depths of 1,000 meters. Pumps have been operated at much greater depths than this so there is no insuperable problem, it is largely a matter of using the correct design which makes allowance for the fact that there will be a considerable hydrostatic pressure in operation. Sea-water pumps are typically about 70% - 80% efficient in converting electric power to mechanical power.

## 4 Economic Considerations.

We should, perhaps, consider the economics of OTEC in the same way as the power industry would assess any new power plant by looking at:

1)      the capital cost per kilowatt of installed power;

2)      the production cost per kilowatt-hour.

The first of these is self explanatory while the second is calculated by dividing the annual outgoings by the annual kilowatt-hours produced. We shall need to remember that the annual outgoings are made up from the annual capital plus interest repayments, the annual operational and maintenance costs and the annual fuel costs. The capital and interest repayments depend very markedly on the time over which the capital has to be repaid and the interest rates involved. There are standard conventions for capital repayment rates which differ from one country to another. In an OTEC plant the convention is that fuel costs are zero but the power produced is the either the surplus power or the gross power depending on the whether the plant is designed for primary electricity production or for the support of mariculture operations. If the production cost is greater than the current market rate for electricity the plant will run at a loss from the start. However, even if the production cost is less than the market rate, so that there will be some profit margin, there will still be concern if the capital cost is significantly higher than for alternative power production techniques.

The early studies, commented on in section 1, indicated that the capital cost of a closed cycle OTEC plant would be very high when compared with that of thermal power stations. The largest capital cost items were the heat exchangers and, in the case of primary power production, the next largest cost was that of the sea-water systems. However, these studies also indicated that the unit costs would become comparable as the size of the OTEC plant was increased and would eventually become competitive if the cost of fuel oil for thermal plants continued to increase. As a result of these considerations the development of closed cycle OTEC virtually stopped when oil prices fell in the early 1980's.

The high cost of the heat exchangers was due to the choice of titanium as the best material to use. Titanium is very resistant to sea-water but it is not made in very large quantities so its cost is high and it is also expensive to fabricate. For these reasons it is an expensive material to use. The sea-water systems are expensive partly because of the material costs and partly because of the installation costs. They are clearly less expensive for a floating platform but floating systems have the additional capital cost of the barge, its mooring system and the transmission cable to take the surplus or net power ashore. However, until closed cycle OTEC is seen to have a competitive unit cost and a more reasonable capital cost it is unlikely to be considered a serious technology. In the next sections we will consider ways of reducing the capital costs of a closed cycle OTEC plant.

## 4.1 Cost Reduction Techniques

### 4.1.1 Heat Exchangers

It will be recalled from table 1 that the important parameter for a heat exchanger is the dimensionless quantity $X$. One would like a large value for $X$ to increase the gross power produced. Now the value of $X$ depends on the thermal conductance times the area of the heat exchanger whereas the cost depends primarily on the area. Thus one important way to reduce the capital cost would be to improve the thermal conductance per unit area.

One way to do this is to increase the sea-water conductance by increasing the water flow velocities over the surface of the heat exchanger. However, this technique also increases the pumping head through the heat exchanger and this reduces the gross power production of the plant, so a balance has to be struck. Another way is to improve both the sea-water conductance and the working fluid conductance by better design of the surfaces themselves. This type of work is part of a major branch of heat and mass transfer studies and a detailed discussion would be outside the scope of this article. Some of the important work in this field is to be found in the following references: in the U.S.A. (23)(24)(25) and in Japan (26)(27)(28)(29). Significant improvements have been made in both the evaporator and the condenser partly by increasing the actual surface area above the nominal area for a smooth surface and partly by improving the thermal conductances themselves. The improvements in some of the individual heat transfer conductances were very striking - in some cases up to 100% - but, at this stage, it is more difficult to estimate the cost reductions that should result by adopting the best of these techniques. In the case of the condenser, Kajikawa estimated that a cost reduction of 30% should be possible (28).

Another possibility for reducing the cost of heat exchangers, first suggested by the Westinghouse study (6), was to see if a less expensive material, such as aluminium, could be used in place of titanium. The cost of titanium, for a given sheet thickness, is about 22 times greater than the cost of aluminium. Thus if aluminium could be used as a satisfactory replacement there was the potential for a very significant cost reduction. The general assumption was that aluminium would not withstand the corrosive effects of sea-water. However, aluminium has been used, from time to time, in ship construction - the Royal Navy's Type 21 frigates have an aluminium superstructure and they are still in active service.

The first experimental work on aluminium was done in 1982 by Sasscer (30) who examined the corrosion properties of diffusion zinc treated and zinc clad aluminium. The next experimental work was by Larsen-Basse (31). A more extensive set of data was reported by workers from the Argonne National Laboratories (20)(21) which indicated that aluminium could be a suitable material for OTEC heat exchangers. Further extensive testing of aluminium has been carried out by Alcan International of Canada. In 1985 Goad, of Alcan's Kingston Laboratories, began a series of systematic corrosion tests on a range of their proprietary aluminium alloys (32). He commissioned an automatic system at the Natural Energy Laboratory at Keahole Point, Hawaii. This chlorinated the sea-water at a low level and was used to examine both samples and welded sections in a long series of corrosion tests on a range of Alcan's proprietary aluminium alloys in both warm and cold sea-water. These tests are still continuing. The main results to date are:

i)      many aluminium alloys show erratic corrosion performance but some show consistently good performance. This indicates that one needs both the right alloy coupled with good quality control in its manufacture. Corrosion is primarily an electrochemical process and is greatly accelerated if there are small differences in chemical composition or even mechanical properties over the surface - hence the need for the good quality control procedures;

ii)     in warm sea-water some aluminium alloys, such as AA5083, show erratic
        corrosion performance but some show consistently good performance (33).
        Simple alloys such as AA1100, AA6063, AA3003 could last for extended
        periods, possibly decades;

iii)    in cold deep sea-water one needs a protective outer layer of zinc to control
        the pitting corrosion. A simple clad layer of a zinc enriched alloy gives
        protection that should last for between 10 and 20 years;

        In summary this work has demonstrated that the right alloys, coupled with
the right quality control in their production, can have a good corrosion resistance
to both warm surface and cold deep sea-water. Other work carried out by Alcan
over a 20 year period indicates that, in general, aluminium corrosion is controlled
primarily by the toughness of the initial protective layer that is formed. If the
aluminium is going to corrode it generally does so in the first 6 to 18 months.
However, if it survives this period it is likely to survive for the next 20 years. This
work opened up the possibility that aluminium could be used for the manufacture
of heat exchangers for OTEC applications.

        The possible corrosive effects of the ammonia on aluminium has also been
studied by Schrieber et Al(34). They found that ammonia and ammonium hydroxide
are not corrosive to aluminium and, in fact, actually somewhat inhibit the corrosion
of aluminium.

        In 1988 GEC and Alcan started work on the design of a range of pilot closed
cycle OTEC plants using these aluminium alloys for the heat exchangers. The first
step in this work was to recognize the limitations of conventional heat exchangers
which had not been optimised specifically for OTEC applications (35). The
limitations with conventional heat exchangers are:

i)      flow back pressures are generally high which increases the pumping head
        losses in the system;

ii)     the method of manufacture is expensive thus largely negating the initial cost
        advantage;

iii)    the manufacture generally involves welding or brazing which would
        introduce corrosion problems for aluminium or be prohibitively expensive
        if cold metal bonding, such as explosive welding, were used.

        This work has resulted in new designs for aluminium heat exchangers are
based on the concept of manifolding together a series of parallel plate elements
whose separation is designed to achieve the required back pressure. The back
pressure can be as low as 2 kPa so that banks of these parallel plate elements can
be immersed in open pools through which the sea-water flows. These heat
exchangers can, if necessary, be mechanically cleaned while in operation. However,
if higher flow rates are used both the back pressure and the heat transfer performance
increase and the open swimming pool arrangement becomes unsatisfactory. In this
case a "containerized" solution is more effective.

        The designs are inherently modular, have virtually no welding or brazing
and make considerable use of Alcan's low cost roll-bonding technique (36) for the
main plate elements. Groups of these plates can be manifolded together to form
modules which can, if necessary, be replaced while the OTEC plant is operating.
This modular design will also make it easier to transport and erect the heat

exchangers at the site. A conservative estimate is that in a pilot plant, of about 1,000 kW mean gross power, the total costs of these new heat exchangers should be less than $900 per installed gross kilowatt.

## 4.1.2 The Expansion Turbine.

As indicated in section 2.5 the unit cost of turbo-generators range from about $2,000/kW for units of about 75 kW to $300/kW for units of 10,000 kW. These turbo-generators have been very highly developed for the process industries where much larger pressure and temperature differences are to be found. As a consequence they are over-designed for OTEC applications and so there is room for further cost reductions. Since one is dealing with small temperature differences and large mass flow rates, the rotation speed of the turbine rotor can be modest, in fact, for the larger gross power outputs, it is at the synchronous speed of the generator so that no gearbox is required. One possibility for reducing costs would be if the rotors could be made by casting rather than by machining. Another is that reciprocating engine expanders could be used in place of expansion turbines especially at the smaller power outputs.

### 4.1.3 Sea-Water Pipes

If the OTEC plant is designed solely for the production of electric power the costs of the sea-water systems become a major cost for the plant. However, if the sea-water systems are designed for a large mariculture operation their costs should not be counted as part of the cost of an auxiliary OTEC plant. Nevertheless any reductions in the costs of sea-water systems help considerably with the economics of large mariculture operations.

The costs of sea-water pipelines for land based systems is inevitably specific to the site. There is a considerable civil engineering effort to install pipes that go through the near-shore surf zone. Costs fall if the pipe work can be fabricated at the site and, for large diameter pipes, laminated ferrocement, which can be fabricated on site, may have something to offer (37). In section 3.1.2 we reviewed some of the ideas for making and installing cold water pipelines. These are usually the most expensive part of the sea-water systems. The technique that, in my view, holds the most promise for reducing costs is that proposed by Wenzel (19) - the "soft pipe." The cost reduction comes from two contributions. Firstly, the reduction in the cost of the pipe material and its fabrication and secondly, the reduction in the cost of deployment of this type of pipeline. I believe that this idea is capable of considerable refinement in the future but, at this stage, there is no experience with the costs of such a pipe.

## 4.2 Co-Products.

In the early 1980's a closed cycle OTEC plant was viewed purely as an alternative way for generating electric power and the economic discussion has been in these terms. However, there has been a very important parallel development taking place at the Natural Energy Laboratory of Hawaii at Keahole Point since 1981 that is not connected with power generation at all. That is the exploitation of deep cold sea-water as a valuable natural resource for aquaculture. Deep ocean water is very rich in nutrients that support marine life. It is also remarkably free of pathogens and, as a consequence, it is ideal for mariculture (38). It is worth noting that about 40% of the world's annual fish catch comes from 0.1% of the oceans and that these areas are those in which this deep, nutrient rich, sea-water is brought to the surface by natural causes. At the Natural Energy Laboratory they simply use sea-water pumps and cold water pipelines for the same purpose. In addition, for some types of aquaculture, warm sea-water is also used. The largest commercial mariculture operator at Keahole point installs his own sea-water pipelines and pumps and buys the electricity he needs from the local grid and, not surprisingly, the power costs are a significant part of his annual outgoings.

The importance of this for a land based OTEC plant is that the pumping of the sea-water, which, for a dedicated power production plant, is a necessary evil, now becomes a primary requirement for such a large mariculture operation. In such cases an auxiliary OTEC plant can provide all the pumping power required plus a useful surplus. The OTEC plant slightly heats the cold sea-water which is generally an advantage for the mariculture operations. In these circumstances the gross power, rather than the surplus or net power becomes the measure of the power output and the capital cost of the sea-water systems is no longer a part of the capital cost of such an auxiliary plant. Thus, as a consequence of this parallel development of deep sea-water aquaculture, the whole perception of what constitutes the capital and the unit costs of a closed cycle OTEC plant change for the better. This will further improve as new uses are found for this deep ocean sea- water.

Other applications for the cold water that have been demonstrated at the Natural Energy Laboratory of Hawaii are its use in horticulture and in direct air conditioning. Its use in horticulture gives the ability to control the temperature in the soil and, if required, immediately above the soil. It also condenses water vapour from the air to supply the plants with fresh water. This results in the rapid growth of crops that could not normally grow in the area. Commercial use of this technique is just beginning. The first experiments in the use of cold sea-water in air conditioning demonstrated an 80% reduction in power costs and its use in the production of potable water is also under consideration (17). As these developments continue and grow the role of an auxiliary closed cycle OTEC plant becomes that of simply being a first user of the sea-water resource.

We hope, with the permission of the Natural Energy Laboratory, to demonstrate an auxiliary closed cycle OTEC plant using our aluminium heat exchangers. If the water supply were sufficient to generate a mean gross output of 500 kW, the unit capital cost would be less than $3,500 per kilowatt - with the heat exchangers contributing less than $1,000/kW to this figure. If one assumes a 20 year pay-back period at a 10% interest rate the unit production cost would be about $0.083 per kilowatt hour.

Thus the possibility that a land based closed cycle plant can be simply a first user of a warm and cold sea-water resource that passes on to other down stream users provides the answer to the question raised in the introduction namely, how can one finance the incremental development of closed cycle OTEC technology if, as the earlier studies indicated, small scale OTEC plants, designed solely for electric power production, would be completely uneconomic? My belief is that these parallel developments of mariculture in Hawaii have now made it possible to resume the development of closed cycle OTEC technology since small OTEC plants can become commercially viable when used as an auxiliary part of a larger multi-product system.

## 5  Environmental Considerations.

There has been an investigation of the possible release of carbon dioxide into the atmosphere from OTEC plants by Green & Guenther (39). Their conclusions were that while there is some release of carbon dioxide from open-cycle OTEC plants, as a result of the sea- water being subjected to a near vacuum, these were only a small percentage of the carbon dioxide emissions from fossil fuelled plants of the same electrical power output. In the case of closed cycle OTEC plant there is an insignificant amount of carbon dioxide released. They also raised the possibility of a very slow release of carbon dioxide if the cold sea-water discharge were made directly into the surface water. However, any such tendency would be completely eliminated if the discharge were made at depths of 100 meters.

Another question that is posed concerns the corrosion products from a closed cycle OTEC plant and their possible environmental impact. A closed cycle OTEC plant heat exchanger with a flow rate of 1 $m^3$/second has $32.5 \times 10^9$ kg/year of sea-water flowing through it. Thus the addition of 32.5 kg/year of corrosion products represents a concentration of one part per billion. In the case of an aluminium heat exchanger designed for this flow rate, 32 kg/year loss of aluminium by corrosion would be a worst case so that there could be an added concentration of aluminium of one part per billion. However, the naturally occurring concentration of aluminium in the sea is 1 part per billion so any environmental impact is likely to be negligible.

Slow leaks of ammonia into the cold water could also pose an environmental problem. The naturally occurring concentration of ammonium ions in deep clod sea-water is 40 parts per billion so that the leak rate needed to reach this figure, in the case just discussed, is 1,300 kg/year or 3.56 kg/day. This is a leak rate that would be easily detected in the make-up circuits of the system.

More serious problems could arise from an accident in which the ammonia was discharged, in a short space of time, into the sea-water. This type of problem is dealt with in the design of a plant. The pressure of the ammonia is continuously monitored as is the make-up rate due to any slow leaks. Any sudden changes in these quantities or any movement outside the design range could be used to start a shut-down of the plant and a rapid by-passing of the sea-water flow around the plant. By ensuring that there is an adequate buffer volume of sea-water in the plant such an accident could be completely contained.

There has also been some speculation that if the nutrient rich deep sea-water were discharged into the surface sea-water it could possibly promote algae blooms which might be regarded as harmful to the environment. Again a discharge at depths of 100 meters would eliminate this possibility. In connection with this it is worth noting that where deep cold water upwells naturally, it has a very beneficial effect on marine life and, as a consequence, on the economies of the neighbouring states. The Japanese are currently conducting experiments to study the beneficial effects of the enhancement of marine bio-productivity from an ocean based system (40). A barge called "The Fertile Sea" is currently operating in the sea of Japan. It has a water spreading system that takes in cold sea-water from a depth of 230 meters, a rate of 0.3 m³/s and a temperature of 4°C and warm sea-water at a rate of 0.6 m³/s and at temperatures ranging from 22 to 25°C. These two flows are mixed together and then sprayed out onto the sea surface through a spray header. The barge also includes a vary small experimental closed cycle OTEC plant of 3.5 kW output. The aim of the mariculture experiment will be to estimate the increase in production of prawns and fish local to the area. If the local fish production does increase dramatically this technique would be an alternative to the Hawaiian procedure of pumping the cold water into ponds on the land. Further, it would open up the possibility for developing floating OTEC plants as part of a larger multi-product system - the electricity produced on the barge being used to pump the sea-water and for other uses.

## 6 Future Developments.

Although the aquaculture developments in Hawaii are less than a decade old their promise seems to be very considerable and, in time, these developments are likely to be replicated in other favourable sites. This should steadily increase the possibilities of small land based closed cycle OTEC plants finding a niche in such developments. Such an auxiliary plant would provide the electricity to operate the sea-water pumps and also produce a useful surplus. This would be seen as a very acceptable arrangement provided the cost of the electricity was competitive with alternative sources.

The economics of the closed cycle OTEC plant improve rapidly as the flow rates increase or if the temperature of the cold water decreases. The gross power output is directly proportional to these flow rates and to the square of the initial temperature difference, while the component costs, especially that of the turbo-generator, begin to benefit from the economies of scale. However, any increase in the flow rates will, for some time, be set by the rate of development of the co-products. In connection with lower cold water temperatures, I understand that the next major pipeline due to be installed in Hawaii will, for the first time, bring up water at 4°C.

While there is still considerable scope for further cost reductions in the heat exchangers for a closed cycle OTEC plant, most of these will arise naturally with experience from fabricating the components. As indicated above there is also considerable scope for cost reductions in the design of turbo-generators by optimising them for closed cycle OTEC. However, the major thrust, for further cost reductions, for a dedicated electricity production system, must lie with the sea-water systems. Here the way forward could well lie with the development of the "soft pipe" concept for the main part of the cold water pipeline. The near-shore pipelines are also expensive and will seem increasingly so if the off-shore part were to become easier. What is required is a combination of strength to withstand heavy wave action with some flexibility to simplify the deployment. Laminated ferrocement may have something to offer here. However, as this part of the system is so site specific it will be less easy to find general techniques.

If one considers the use of off-shore plant these introduce a number of additional problems although they greatly simplify the sea-water systems. I think that if the Japanese "Fertile Sea" experiments prove to be a success this could open up the possibility that here too an OTEC plant could become an auxiliary source of power to drive the sea-water pumps. At present floating systems do not have the this advantage.

Just as the parallel development of deep ocean water aquaculture at the Natural Energy Laboratory of Hawaii has benefited land based OTEC, other parallel developments concerned with utilising waste heat in the power and process industries, may also benefit closed cycle OTEC by steadily improving the components that will be needed. The prospects for closed cycle OTEC are, I believe, better now than they were in the early 1980's, but it will continue to need a good deal of effort - and above all luck - if it is to realise its potential.

# 7 Acknowledgements

In producing this article I have had advice and help from a number of friends and colleagues. In particular I should like to acknowledge the help I have received from Mr Don Lennard, Ocean Thermal Energy Conversion Systems Ltd., Orpington, Kent, UK., Dr Geof Williams, Engineering Research Centre, Whetstone, Leicester, U.K., Dr Joe Van Ryzin, Makai Ocean Engineering, Hawaii, Dr Nigel Fitzpatrick, Alcan International Ltd., Kingston, Ontario and Dr Takenobu Kajikawa of the Electrotechnical Laboratory, MITI, Japan. I would also like to acknowledge my indebtedness to Dr John Craven, of the University of Hawaii, who first inspired me to take the subject of closed cycle OTEC seriously.

# 8 References

1) "International Working Group on Ocean Thermal Energy Systems." June 18th 1986.

2) d'Arsonval,A. Revue scient., pp370-372, 1881.

3) Claude, G., "Power From the Tropical Seas." Mechanical Engineering, Vol. 52, No. 12, pp 1039-1044, 1930.

4) Lockheed Missiles & Space Company, "OTEC Power System Development - Phase I - Final Report" December 1978

5) TRW, Inc., "OTEC Power System Development, Preliminary Design Report, Final, December 1978

6) Westinghouse Electric Corporation, "Power System Development - Preliminary Design, Final Report", December 1978

7) Gritton, E.C., Pei, R.Y., Aroesty, J., Balaban, M.M., Gazley, C., Hess, R.W., & Kras, W.H., "A Quantitative Evaluation of Closed-Cycle Ocean Thermal Energy Technology in Central Station Applications." R-2595-DOE, The Rand Corporation, Santa Monica, California. (1980)

8) "Report of the Eurocean Marine Energy Study Group on the Present and Future Possibilities of Energy Production from Marine Sources" Eurocean, Monaco, (1977)

9) IMFREMER "Ocean Thermal Energy Conversion, the French Programme A Pilot electric Power Plant for the French Polynesia." 1978-82 (1985)

10) OTEC Systems Ltd "A 10 MW Floating OTEC Demonstration Plant Phase 1: First Progress Report." (1984)

11) Trimble, L.C., & Owens, W.L., "Review of Mini-OTEC Performance." 15th IECEC (1980)

12) Douglas, R.H., "OTEC: Solar Energy from the Sea, Quest", TRW DSSG (1979)

13) Lorenz, J.J., Yung, D., Howard, P.A. & Panchal, C.B. Proceedings of the Eighth Ocean Energy Conference, 1981.

14) Mitsui, T., Ito, F., Seya Y., & Nakamoto, Y., "Outline of the 100 kW OTEC pilot plant in the Republic of Nauru." IEEE/PES Winter Meeting, (1983).

15) Suganuma, T., "Outline of the Tokunoshima OTEC Test Plant." Ocean Development News 11, 2, 37-40 (1983)

16) Kamogawa, H., "OTEC Research in Japan." Energy, Vol5, pp481-492, (1980)

17) Johnson, F.A., "Energy from the Oceans: A Small Land Based Ocean Thermal Energy Plant". EEZ Resources Technology Assessment Conference, Hawaii, 22-26 January pp 2-20, (1989)

18) Van Ryzin, J. Makai Ocean Engineering, Inc. at Waimanalo, Hawaii. Private Communication

19) Bottom-Mounted OTEC Sea-water Systems Workshop. Vols 1 & 2, April 1988 published by the Pacific International Centre for High Technology Research, Honolulu, Hawaii.

20) Panchal,C.B., "Experimental Investigation of Marine Biofouling and Corrosion for Tropical Sea-water" paper at conference "Advances in Fouling Science and Technology" Portugal p7 (1987)

21) Thomas, A., & Willis, D.L., "Biofouling and Corrosion Resistance for Marine Heat Exchangers." pp 38 - 41, Oceans 89.

22) Panchal,C.B., "Experimental Investigation of Sea-water Biofouling for Enhanced Surfaces" ASME/AIChE National Heat Transfer Conference, p26 (1989)

23) Bergles,A.E. & Webb,R.L., "Augmentation of Heat and Mass Transfer" published by Hemisphere Publishing, New York, NY (1986 ).

24) Panchall,C.B. & France,D.M., "Performance Tests of the Spirally Fluted Tube for Industrial Co-generation Applications" Argonne National Lab. Report ANL/CNSV-59 (1986)

25) Webb,R.L., "Enhanced Tubes for Electricity Utility Steam Condensers", paper at conference EPRI Condenser Technology Conference May p 30 (1990)

26) Kajikawa, T., Takazawa, H., & Mizuki, M., "Heat Transfer Performance of Metal Fibre Sintered Surfaces", Heat Transfer Engineering, vol 4, pp57-66, 1983

27)    Kajikawa, T., Takazawa, H., Murata, A., & Nishiyama, K., "OTEC Power System Simulator", Proceedings of Pacific Congress on Marine Technology, Honolulu, Hawaii April 1984

28)    Takazawa, H., & Kajikawa, T., "Condensing Heat Transfer Enhancement on Vertical Spiral Double Fin Tubes with Drainage Gutters", Journal of Solar Energy vol 107, August 1985.

29)    Mori,Y. & Echigo,R, "Heat Transfer Bibliography: Japanese Work" International Journal of Heat and Mass Transfer 30, 417-26 (1987)

30)    Sasscer, D.S., "In-Situ Sea-water Corrosion of Bare Diffusion Zinc Treated and Alclad Aluminium Heat Exchanger Materials." Oceans '82, pp 578-586.

31)    Larsen-Basse, J., "Performance of OTEC Heat Exchanger Materials in Tropical Sea-waters". J. Metals 37, 24 (1985)

32)    Goad, D.G.W., "Corrosion of Aluminium Alloys for OTEC Heat Exchangers". Abstract 279, Electrochemical Society, Honolulu, October 1987.

33)    Schrieber, C.F., Grimes, W.D. & McIlhenny, W.F., "A Study of the Corrosive Effect on Aluminium and CP Titanium of Mixtures of Ammonia and Sea-water that may be encountered in OTEC Heat Exchangers" Final Report, ANL OTEC-BCM-004 Argonne National Laboratory, March 1979.

34)    Fitzpatrick, N.P., Hron, V., Hay, E. & Johnson, F.A. "Alcan's Ocean Thermal Energy Program" 9th International Congress on Energy & Environment, December 1989.

35)    Lennard, D.E. & Johnson, F.A., "British OTEC Programmes: 10 MW Floating and 0.5 MW Land Based". Oceans '88 p1034.

36)    Fulton, C.W., "Continuous Roll-Bonding: A Means of Manufacturing Heat Exchanger Parts and Components". International SAE Congress, Detroit, March 1983, Paper SAE 830023.

37)    Iorns, M.E., "OTEC Sea-water Pipe Cost Comparisons", Proceedings of the International Conference on Ocean Energy Recovery, Honolulu, Hawaii, pp297-306, 1989.

38)    Daniel, T.H., "Aquaculture Using Cold OTEC Water". Proc. Oceans '85 Conference, pp 1284-1289.

39)    Green, H.J., & Guenther, P.R., "Carbon Dioxide Release from OTEC Cycles", Proceedings of the International Conference on Ocean Energy Recovery, Honolulu, Hawaii, pp348-357 1989.

40)    Kajikawa, T., "Status on OTEC Technology Research and Development in Japan", Conference record, 16th Meeting of UJNR MFP pp111-121 (1989).

41)    Mark's Standard Handbook for Mechanical Engineers, Eighth Ed. sections 4-33 to 4-35, McGraw Hill Book Company

# 9 Appendix: Thermodynamics.

## 9.1 Heat Transfer.

Let $\dot{M}_{ew}$ be the mass flow rate of warm sea-water through the evaporator and $\dot{M}_{cw}$ be the mass flow rate of cold sea-water through the condenser then *the rate of absorption of thermal energy by the evaporator*, $\dot{Q}_{er}$, becomes

$$\dot{Q}_{er} = \Delta T_e\, G_e \qquad\qquad 9.1.1$$

where

$$\Delta T_e = T_{ei} - T_{ef} \qquad\qquad 9.1.2$$

and

$$G_e = C_{pw}\, \dot{M}_{ew} \qquad\qquad 9.1.3$$

and where $C_{pw}$ is the specific heat of sea-water and $T_{ei}$ and $T_{ef}$ are the initial and final temperatures of the sea-water entering and leaving the evaporator. Similarly *the rate of release of thermal energy by the condenser*, $\dot{Q}_{cr}$, is defined by the equations:

$$\dot{Q}_{cr} = \Delta T_c\, G_c$$

$$\Delta T_c = T_{cf} - T_{ci}$$

$$G_c = C_{pw}\, \dot{M}_{cw} \qquad\qquad 9.1.4$$

and where $T_{ci}$ and $T_{cf}$ are the initial and final temperatures of the sea-water entering and leaving the condenser - note that $T_{cf}$ is greater than $T_{ci}$. Further, to transfer this heat to the working fluid vapour in the evaporator, the temperature of the vapour, $T_e$, must be less than $T_{ef}$ thus:

$$T_e = T_{ef} - T_{ea}$$

$$= T_{ei} - \Delta T_e - T_{ea} \qquad\qquad 9.1.5$$

where $T_{ea}$ is the terminal temperature difference between the sea-water and the working fluid leaving the evaporator. Similarly to transfer this heat from the working fluid in the condenser to the sea-water the temperature of the vapour, $T_c$, must be greater than $T_{cf}$ thus:

$$T_c = T_{cf} + T_{ca}$$

$$= T_{ci} + \Delta T_c + T_{ca} \qquad\qquad 9.1.6$$

where $T_{ca}$ is the terminal temperature difference between the sea-water and the working fluid leaving the condenser. Thus it follows that:

$$(T_e - T_c) = (T_{ei} - \Delta T_e - T_{ea}) - (T_{ci} + \Delta T_c + T_{ca})$$

$$= \Delta T_i - (\Delta T_e + T_{ea}) - (\Delta T_c + T_{ca}) \qquad 9.1.7$$

where

$$\Delta T_i = T_{ei} - T_{ci} \qquad 9.1.8$$

Note that $\Delta T_i$ is the initial temperature difference available for the OTEC conversion process. Thus it follows that the temperature difference across the heat engine is less than the initial temperature due to these irreversible heat transfer processes.

These five temperature differences, $\Delta T_e$, $T_{ea}$, $(T_e - T_c)$, $T_{ca}$ and $\Delta T_c$ form a temperature "ladder" between the intake temperatures of warm and cold sea-water. In the next sections we show that each is proportional to the initial temperature difference $\Delta T_i$.

## 9.2 Heat Exchangers.

The rate of transfer of thermal energy for the evaporator is given by the equation

$$\dot{Q}_{er} = U_e A_e \Delta\Theta_e \qquad 9.2.1$$

where $U_e$ is the mean thermal conductance, or the overall heat transfer coefficient of the evaporator, $A_e$ is the area of the evaporator surface and $\Delta\Theta_e$ is the *log mean temperature difference* for the evaporator and is defined by the equation:

$$\Delta\Theta_e = \Delta T_e / \text{Ln}[(\Delta T_e + T_{ea})/T_{ea})]$$

$$= \dot{Q}_{er}/(U_e A_e)$$

$$= (\Delta T_e G_e)/(U_e A_e) \qquad 9.2.2$$

where $Ln(x)$ is the natural logarithm of $x$ and where the second step follows from 9.1.1. Thus it follows that:

$$U_e A_e / G_e = \text{Ln}[(\Delta T_e + T_{ea})/T_{ea}]$$

$$= X_e \qquad 9.2.3$$

and that

$$T_{ea} = (Y_e - 1)\Delta T_e \qquad 9.2.4$$

where

$$Y_e = 1/[1 - \exp(-X_e)] \qquad 9.2.5$$

The corresponding equations for the condenser are:

$$T_{ca} = (Y_c - 1)\,\Delta T_c$$

$$Y_c = 1 / [1 - \exp(-X_c)]$$

$$X_c = U_c\,A_c / G_c \qquad\qquad 9.2.6$$

These equations define the values of $T_{ea}$ and $T_{ca}$ when the remaining variables are known. It follows that equation 9.1.7 can be written as:

$$(T_e - T_c) = \Delta T_i - Y_e\,\Delta T_e - Y_c\,\Delta T_c \qquad\qquad 9.2.7$$

## 9.3 The Gross Power Output.

The mechanical gross power produced, $P$, by a reversible heat engine is the product of the heat taken up by the evaporator and the Carnot efficiency and is given by

$$P = G_e\,\Delta T_e\,(T_e - T_c)/T_e \qquad\qquad 9.3.1$$

Now we see from equations 9.1.5, 9.2.4 & 9.2.7 that

$$P = \frac{G_e\,\Delta T_e\,[\,\Delta T_i - Y_e\,\Delta T_e - Y_c\,\Delta T_c\,]}{[\,T_{ei} - Y_e\,\Delta T_e\,]} \qquad\qquad 9.3.2$$

$$= G_e\,\Delta T_e - G_c\,\Delta T_c \qquad\qquad 9.3.3$$

where the second step follows from the first law of thermodynamics. On combining these equations to eliminate $\Delta T_c$ we obtain:

$$\frac{P}{G_e} = \frac{\Delta T_e\,[\,\Delta T_i - \Delta T_e\,(Y_e + Y_c\,G_e/G_c)\,]}{[\,T_{ei} - \Delta T_e\,(Y_e + Y_c\,G_e/G_c)\,]} \qquad\qquad 9.3.4$$

It follows that the gross power is a maximum when:

$$\Delta T_e\,(Y_e + Y_c\,G_e/G_c) = T_{ei}\,[\,1 - \sqrt{(1 - \Delta T_i/T_{ei})}\,]$$

$$= \tfrac{1}{2}\Delta T_i\left[\,1 + \tfrac{1}{4}C_o\,(1+\delta)\,\right] \qquad\qquad 9.3.5$$

where $C_o$ is the theoretical maximum Carnot efficiency defined by the equation

$$C_o = \Delta T_i / T_{ei} \qquad\qquad 9.3.6$$

and

$$\delta = \tfrac{1}{2}C_o\left[\,1 + \tfrac{5}{8}C_o + \tfrac{5.7}{8.10}C_o^2 + ..\right] \qquad\qquad 9.3.7$$

The maximum gross power, $P_m$ is given by the equation:

$$P_m = \frac{\Delta T_i^2\,(1+\delta)}{4\,(Y_e/G_e + Y_c/G_c)\,T_{ei}} \qquad\qquad 9.3.8$$

It also follows that the temperature drop in the condenser is given by:

$$\Delta T_c \left( Y_e + Y_c\, G_e / G_c \right) \;=\; \tfrac{1}{2}\Delta T_i \left( G_e / G_c \right)\left[\, 1 - \tfrac{1}{4}C_o\,(1+\delta)\,\right] \qquad\qquad 9.3.9$$

and that the temperature drop across the reversible heat engine is given by:

$$T_e - T_c \;=\; \tfrac{1}{2}\Delta T_i\,[\,1 - C_o\,F_{ec}\,(1+\delta)\,] \qquad\qquad 9.3.10$$

where

$$F_{ec} \;=\; \frac{[\,Y_e / G_e - Y_c / G_c\,]}{[\,Y_e / G_e + Y_c / G_c\,]} \qquad\qquad 9.3.11$$

Finally it follows that the conversion efficiency for the complete system, $C_1$, is given by:

$$C_1 \;=\; \tfrac{1}{2}C_o\left[\, 1 + \tfrac{1}{4}C_o\,(1+\delta)\,\right] \qquad\qquad 9.3.12$$

These equations demonstrate that all five temperature drops in the "ladder" - see section 9.1 - are proportional to the initial temperature difference $\Delta T_i$. They also demonstrate that the maximum gross power output occurs when just under half the initial temperature difference appears across the heat engine and the remainder is used in the irreversible processes of driving heat through the heat exchangers. It is also seen that the efficiency of the cycle is just over half the theoretical Carnot efficiency. The effect of the heat exchangers on the maximum gross power is also evident since as the values of $Y_e$ or $Y_c$ increase the gross power decreases.

In table 4 an example is given of the maximum gross power output for a closed cycle OTEC plant with $X_e = X_c = 1.0$ and with sea-water flows of one cubic meter per second for each heat exchanger (15,856 US gals/min or 13,198 Imp gals/min).

Note that the temperature drop across the heat engine, 10.999°K, is very close to half the initial temperature drop of 22.0°K. As a consequence the conversion efficiency is reduced from a theoretical value of 7.33% to 3.74%.

The effect of changing the performance of the heat exchangers in the previous example was shown in table 1 (see section 2.2) above where the maximum gross power output, calculated exactly, was shown for a series of values of $X_e$ which, in each case, is taken to be the same as $X_c$.

Calculation of the Maximum Gross Power Output for a Reversible Heat Engine with one cubic metre per second Flow Rates.

| | | | | |
|---|---|---|---|---|
| $X_e =$ | 1.0 | | $X_c =$ | 1.0 |
| $Y_e =$ | 1.582 | | $Y_c =$ | 1.582 |
| $T_{ei} =$ | 300 °K | | $T_{ci} =$ | 278 °K |
| $C_{pw} =$ | 3.993 kW/kg/°K | | $\Delta T_i =$ | 22 °K |
| $\dot{M}_e =$ | 1,025 kg/s | | $\dot{M}_c =$ | 1,030 kg/s |
| $G_e =$ | 4,093 k/°K | | $G_c =$ | 4,113 kW/°K |
| $\Delta T_e =$ | 3.551 °K | | $\Delta T_c =$ | 3.439 °K |
| $T_e - T_c =$ | 10.999 °K | | $\delta =$ | 0.0384 |
| $C_0 =$ | 7.33 % | | $C_1 =$ | 3.74 % |
| $P_m =$ | 543.1 kW | | $\dot{Q}_{er} =$ | 14,536 kW |

Table 4.

## 9.4 The Rankine Cycle.

### 9.4.1 The Idealised Case.

In this ideal case, both the expansion of the saturated vapour through the turbine and the compression of the liquid by the feed pump is isentropic. There are four steps in the Rankine cycle.

1) At the output of the evaporator saturated vapour is supplied to the expansion turbine with the following thermodynamic properties:

    temperature  $T_e$          enthalpy    $H_{ge}$
    pressure     $P_e$          entropy     $S_{ge}$

2) At the input to the condenser, i.e. after the isentropic expansion through the turbine, the, by now, wet vapour has the following thermodynamic properties:

    temperature  $T_c$          enthalpy    $H_{wgc}$
    pressure     $P_c$          entropy     $S_{ge}$

3) At the input to the feed pump, i.e. after the wet vapour has condensed into a liquid, the working fluid has the following thermodynamic properties:

    temperature  $T_c$          enthalpy    $H_{fc}$
    pressure     $P_c$          entropy     $S_{fc}$
                                volume      $V_{fc}$

4)    Finally at the input to the evaporator, i.e. after the isentropic compression of the, virtually incompressible liquid, by the feed pump, the working fluid has the following thermodynamic properties:

| temperature | $T_c$ | enthalpy | $H_{cfc}$ |
| pressure | $P_e$ | entropy | $S_{fc}$ |
| | | volume | $V_{fc}$ |

5)    The fluid is then re-heated and boiled in the evaporator to re-emerge from its output with the thermodynamic properties listed in step 1 above.

One can now write down the quantitative performance of this cycle, for *each unit mass of the working fluid circulating around the system*, as follows:

1)    the heat supplied by the evaporator, $Q_e$ , is given by

$$Q_e \;=\; H_{ge} - H_{cfc} \qquad\qquad 9.4.1$$

2)    the work generated by an ideal isentropic turbine, $W_s$, is given by

$$W_s \;=\; H_{ge} - H_{wgc} \qquad\qquad 9.4.2$$

3)    the heat removed by the condenser, $Q_c$, is given by

$$Q_c \;=\; H_{wgc} - H_{fc} \qquad\qquad 9.4.3$$

4)    the work absorbed by an ideal isentropic feed pump, $W_f$, is given by

$$W_f \;=\; H_{cfc} - H_{fc} \qquad\qquad 9.4.4$$

thus it follows that
$$\begin{aligned} Q_e + W_f \;&=\; H_{ge} - H_{fc} \\ &=\; Q_c + W_s \qquad\qquad 9.4.5 \end{aligned}$$

and that the net work, which in OTEC terminology, is called the gross work produced by the cycle, $W_{gross}$ , is

$$W_{gross} \;=\; W_s - W_f \qquad\qquad 9.4.6$$

All that remains is the definitions for the enthalpies appearing in these four equations. Now $H_{ge}$ and $H_{fc}$ are standard thermodynamic properties defined above. These can be read from standard thermodynamic tables (41). The enthalpy of the compressed fluid from the condenser, $H_{cfc}$ , can be calculated from equation 9.4.4 by noting that the work absorbed by the feed pump is the difference between the product of the pressure difference between the evaporator and the condenser and the specific volume of the fluid at $T_c$ . Thus

$$W_f \;=\; ( P_e - P_c ) V_{fc} \qquad\qquad 9.4.7$$

and, since all the quantities on the right of equation 9.4.7 can be read from standard thermodynamic tables, this defines $H_{cfc}$ .

Since the expansion of the saturated vapour through the turbine is isentropic, then the entropy of the vapour as it leaves the turbine is $S_{ge}$ . During the expansion the vapour cools and some of it condenses to a liquid, releasing its latent heat in doing so. Thus the entropy of the wet vapour will lie somewhere between that of saturated vapour and saturated liquid - which is significantly lower - at the condenser temperature. One defines a dryness fraction, $D_f$ , for the wet vapour as

$$S_{ge} = D_f S_{gc} + ( 1 - D_f ) S_{fc} \qquad 9.4.8$$

Similarly the enthalpy of the wet vapour, $H_{wgc}$ , will also lie between the enthalpy of saturated vapour, $H_{gc}$ , and saturated liquid, $H_{fc}$ , at the condenser temperature so that

$$
\begin{aligned}
H_{wgc} &= D_f H_{gc} + ( 1 - D_f ) H_{fc} \\
&= H_{gc} - ( 1 - D_f ) ( H_{gc} - H_{fc} ) \\
&= H_{gc} - ( 1 - D_f ) T_c ( S_{gc} - S_{fc} ) \\
&= H_{gc} - T_c ( S_{gc} - S_{ge} ) \qquad 9.4.9
\end{aligned}
$$

Since all the terms on the right hand side of equations 9.4.9 can be read from standard thermodynamic tables, this defines $H_{wgc}$ .

The mass flow rate of the working fluid, $\dot{M}_a$ , is defined by the equation

$$
\begin{aligned}
\dot{M}_a &= \dot{Q}_{er} / Q_e \\
&= G_e \, \Delta T_e / Q_e \qquad 9.4.10
\end{aligned}
$$

where $\dot{Q}_{er}$ is the thermal power supplied to the evaporator from the sea-water. It readily follows that the power produced by the turbine, $P_s$ ,is given by:

$$P_s = W_s \dot{M}_a \qquad 9.4.11$$

the power absorbed by the feed pump, $P_f$ is given by:

$$P_f = W_f \dot{M}_a \qquad 9.4.12$$

and the gross power produced, $P_{gross}$ , is given by:

$$P_{gross} = W_{gross} \dot{M}_a \qquad 9.4.13$$

An important criterion for assessing thermodynamic cycles sensitivity to irreversible processes is the *work ratio*, $R_w$ , which is ratio of the net work (or gross work in OTEC terminology) to the positive work done in the cycle. In the case of the ideal Rankine cycle this is given by

Calculation of the Maximum Gross Power Output for an Ideal
Rankine Cycle with Ammonia using the Input Data from Table 4.

| | | | | |
|---|---|---|---|---|
| $T_e =$ | 294.3 °K | | $T_c =$ | 283.4 °K |
| $P_e =$ | 889.4 kPa | | $P_c =$ | 620.9 kPa |
| $H_{ge} =$ | 1,463 kJ/kg | | $H_{gc} =$ | 1,454 kJ/kg |
| $S_{ge} =$ | 5.082 kJ/kg/°K | | $S_{gc} =$ | 5.210 kJ/kg/°K |
| $H_{fe} =$ | 280.6 kJ/kg | | $H_{fc} =$ | 229.1 kJ/kg |
| $S_{fe} =$ | 1.063 kJ/kg/°K | | $S_{fc} =$ | 0.886 kJ/kg/°K |
| $T_e - T_c =$ | 10.88 °K | | $V_{fc} =$ | 1.602 l/kg |
| $W_f =$ | 0.4300 kJ/kg | | $\dot{Q}_{er} =$ | 14,675 kW |
| $H_{cfc} =$ | 229.6 kJ/kg | | $\dot{M}_a =$ | 11.90 kg/s |
| $D_f =$ | 97.0 % | | $P_s =$ | 537.7 kW |
| $H_{wgc} =$ | 1,418 kJ/kg | | $P_f =$ | 5.12 kW |
| $W_s =$ | 45.21 kJ/kg | | $P_{gross} =$ | 532.6 kW |
| $Q_e =$ | 1,234 kJ/kg | | $R_w =$ | 99.1 % |
| $Q_c =$ | 1,189 kJ/kg | | $C_I =$ | 3.63 % |

Table 5.

$$R_w = P_{gross} / P_s \qquad\qquad 9.4.14$$

Table 5 gives the results for a Rankine cycle using the same input data as in
table 4 but now using ammonia as the working fluid. Note that the gross power is
533 kW slightly lower than the theoretical gross power of 543 kW for a reversible
heat engine. Note too that the temperature drop across the expansion turbine is
reduced to 10.88°K.

### 9.4.2 The Actual Case.

Actual expansion turbines are not isentropic so that the actual work
produced, $W_{as}$, is less than the theoretical amount, $W_s$. One defines an isentropic
efficiency, $E_{eff}$, for an expansion turbine as

$$W_{as} = E_{eff} W_s \qquad\qquad 9.4.15$$

and it follows that the actual enthalpy of the wet exhaust vapour, $H_{awgc}$, is greater
than that in the ideal case, $H_{wgc}$, and is given by

$$H_{awgc} = E_{eff} H_{wgc} + (1 - E_{eff}) H_{ge} \qquad\qquad 9.4.16$$

Note that as the isentropic efficiency tends to zero, the decrease in enthalpy of the vapour passing through the turbine, also tends to zero. It will also be seen from equations 9.4.13 and 9.4.2 that

$$W_{as} = E_{eff} W_s$$

$$= E_{eff} H_{ge} - E_{eff} H_{wgc}$$

$$= H_{ge} - [ E_{eff} H_{wgc} + ( 1 - E_{eff} ) H_{ge} ]$$

$$= H_{ge} - H_{awgc} \qquad 9.4.17$$

The heat absorbed by the condenser, $Q_{ac}$ , is now given by

$$Q_{ac} = H_{awgc} - H_{fc} \qquad 9.4.18$$

Note that $Q_{ac}$ is larger than $Q_c$ since $H_{awgc}$ is greater than $H_{wgc}$ . It will also be seen that as before

$$Q_e + W_f = Q_{ac} + W_{as} \qquad 9.4.19$$

Table 6 lists the modifications that result if an 85% isentropic expansion efficiency is assumed for the turbine. Note the increase in the condenser duty, 1,189 kJ/kg to 1,196 kJ/kg, and the reductions in the conversion efficiency - 3.63% to 3.07% - the work ratio - 99.1% to 98.9% - and the gross power - 533 kW to 451 kW.

The Effect of an 85% Expansion Efficiency
on the Results from Table 5.

| | | | |
|---|---|---|---|
| $T_e =$ | 294.3 °K | $T_c =$ | 283.5 °K |
| $P_e =$ | 889.6 kPa | $P_c =$ | 621.4 kPa |
| $H_{ge} =$ | 1,463 kJ/kg | $H_{gc} =$ | 1,454 kJ/kg |
| $S_{ge} =$ | 5.082 kJ/kg/°K | $S_{gc} =$ | 5.210 kJ/kg/°K |
| $H_{fe} =$ | 280.6 kJ/kg | $H_{fc} =$ | 229.3 kJ/kg |
| $S_{fe} =$ | 1.063 kJ/kg/°K | $S_{fc} =$ | 0.886 kJ/kg/°K |
| $T_e - T_c =$ | 10.87 °K | $V_{fc} =$ | 1.602 l/kg |
| $W_f =$ | 0.4296 kJ/kg | $\dot{Q}_{er} =$ | 14,654 kW |
| $H_{cfc} =$ | 229.7 kJ/kg | $\dot{M}_a =$ | 11.88 kg/s |
| $D_f =$ | 97.0% | $P_s =$ | 455.8 kW |
| $H_{awgc} =$ | 1,425 kJ/kg | $P_f =$ | 5.10 kW |
| $W_s =$ | 38.37 kJ/kg | $P_{gross} =$ | 450.7 kW |
| $Q_e =$ | 1,234 kJ/kg | $R_w =$ | 98.9% |
| $Q_c =$ | 1,196 kJ/kg | $C_1 =$ | 3.07% |

Table 6.

## 9.5 Hydraulics

As is well known the resistance to the flow of water through a pipe is approximately proportional to the square of the velocity and to the wetted area. By equating this to the suction head times the cross section area of the pipe one arrives at the simple formula:

$$H_s = A_o (L / \rho) (Q^2 / d^5) \qquad 9.5.1$$

where $H_s$ is the suction head, $A_o$ is a constant for the particular type of pipe, $(L / \rho)$ is the length of the pipe divided by the density of the water, $Q$ is the water volume flow rate and $d$ is the diameter of the pipe. Empirical formula are usually more accurate than 9.5.1 and one example, the Hazen and Williams formula, is given by:

$$H_s = B_o (L / \rho) (Q^{1.85} / d^{4.86}) \qquad 9.5.2$$

In addition to the suction head there is, of course, an entry and exit head due to the initial acceleration and subsequent retardation of the water. This entry and exit head is given by the equation:

$$H_{ex} = K (v / a_o)^2 / (2 g) \qquad 9.5.3$$

where $H_{ex}$ is the entry and exit head, $(v / a_o)$ is the flow velocity divided by the cross section area, $g$ is the acceleration of gravity and $K$ is a constant whose value is typically 1.5.

To these pumping heads, associated with the sea-water pipelines, must be added any lift above sea level, the pumping heads generated by the flow resistance in the heat exchangers and, in the case of the cold water, a density head due the difference in density of the cold and warm sea-water. If the cold sea-water is at a temperature of 4°C this density head is about 1.2 meters. The mechanical pumping power for the sea-water can then be calculated from the equation:

$$P_{pump} = Q \rho g H_{total} \qquad 9.5.4$$

where $H_{total}$ is the total pumping head.

The temperature rise, per meter length of pipe, is given by the heat balance equation:

$$Q \rho C_w dT = \pi U d \Delta T \qquad 9.5.5$$

where the left of this equation is the product of the mass flow rate, the specific heat and the temperature increase of the sea-water while the right of this equation is the product of the thermal conductance of the pipe material, the perimeter of the pipe and the temperature difference between the water in the pipe and the sea-water outside. The total temperature change is found by integrating this equation over the length of the pipe knowing the temperature profile of the sea. Clearly to maintain a small temperature rise the thermal conductance should be small. However, if the mass flow rate is kept constant, one sees that the temperature rise becomes proportional to the pipe diameter.

# CHAPTER 6

## STATE OF THE ART IN OPEN-CYCLE

## OCEAN THERMAL ENERGY CONVERSION

By

Michel Gauthier[1], Jean Marvaldi[2], Federica Zangrando[3]

1/   Institut Français de Recherche pour l'Exploitation de la Mer (IFREMER)
Delegation in New Caledonia
Quai des Scientifiques, BP 2059
Noumea, Nouvelle Caledonie

2/   IFREMER,Center of Brest
BP 70
29280 Plouzané, France

3/   National Renewable Energy
Laboratory (NREL)
(Formerly the Solar Energy
Research Institute - SERI)
1617 Cole Blvd.
Golden, CO. 80401, USA

## Table of contents

Abstract


The concept of the open-cycle ocean thermal energy conversion (OC-OTEC) was first proposed and demonstrated by the french industrial manager Georges Claude around 1930.
Some projects of OC-OTEC plants were developed by France until the 1950s but none reached the stage of construction.
A period of inactivity in that field followed until the 1973 oil crisis. Then the USA and french governments funded multiyear programs aiming at the reassessment of OC-OTEC.
The french program culminated in 1986 with the design of a 5 MW net power plant for Tahiti Island, which was finally abandoned.
The US program, after extensive tests on OC-OTEC components, resulted in a 210 kW gross power system currently under construction in Hawaii.

In the OC-OTEC process, less than one percent of the warm seawater flow is vaporized in a vacuum chamber. The steam flow is then directed through a turbine to a condenser refrigerated by a flow of cold deep seawater.
The small available temperature difference between warm and cold water flows can only give a maximum theoretical efficiency of the process of some percents. Thus superior performances are required from all the system components.
From the tests of various evaporator concepts, the vertical spout evaporator emerged as the most efficient one. A predeaeration chamber may be incorporated to reduce the quantity of noncondensable gases evolving from the warm water in the evaporator.
In direct-contact condensers, steam is condensed on jets or films of cold seawater. The geometry is designed to maximize the contact area between the fluids while minimizing head losses and the quantity of steam necessarily extracted with the noncondensable gases. Tube-bundle surface condensers can also be used to recover fresh water as a by-product at the cost of a small decrease of thermal efficiency.
Due to the large specific volume of the low pressure steam, the size of the turbine, with existing technology, limits the unit power of a modular OC-OTEC system to a few megawatts.
An efficient multi-stage venting system is necessary for the recompression of the noncondensables gases extracted from the condenser.
Thermo-hydraulic characteristics of the components can be described by mathematical models, incorporating experimental results.
Component models can be integrated with additional cost equations in simulation computer programs to perform a sensitivity analysis or cost optimization of a complete OC-OTEC system.

Besides the availability of fresh water as a by-product in an OC-OTEC electrical plant equipped with surface condensers, the ocean temperature gradient can also be used for the unique purpose of freshwater production in various multi-stage desalination system with the advantage of reduced energy consumption.

## 6.0.    Introduction

Historically the first idea that emerged for harnessing ocean thermal energy was that of the closed cycle, conceived by Arsène d'Arsonval in 1881. In the closed-cycle process, the warm-water flow provides and the cold water flow removes heat energy from a chemical working fluid. The vapor of that fluid generates mechanical energy in a turbine. Heat-conveying water flows and the working fluid have to be separated by watertight heat exchange surfaces, as detailed in Chapter 5.

Some forty years after the emergence of the closed-cycle idea, a new concept was presented by the French engineer and industrial manager Georges Claude [5]. Heat transfer from warm water could be achieved by vaporizing a fraction of that flow in a partial vacuum, and the steam so produced could be used as the working fluid. Although that fluid was at a relatively low temperature and saturation pressure, it was recognized that its energy content was appreciable and could be recovered in a turbine. Working-fluid generation from the warm-water flow could be achieved in direct-contact heat exchangers, eliminating watertight barriers. Heat exchangers should be a lot simpler and cheaper than those required for the closed cycle. Inversely, the heat exchangers must be maintained under vacuum, and the very high specific volume of low-pressure steam requires larger turbines and other process components. A block diagram of the open, or Claude cycle is shown in Figure 6.1 and details are provided in Section 6.2.1.

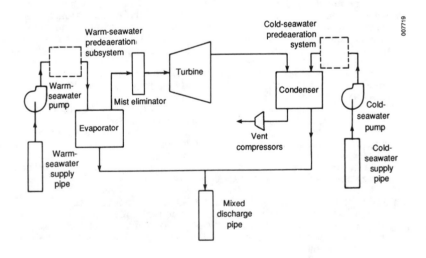

Figure 6.1 : Block diagram of open-cycle
(Claude cycle) OTEC system

Open-cycle research is based on the projected advantages of completely or partially eliminating conventional heat exchangers in the system and the penalties of performance and cost associated with them. Whether the full potential for this approach can be realized depends on achieving acceptable levels of power-conversion efficiency and on reducing parasitic scale.

## 6.1. Historical Development

Georges Claude, the promoter of the open-cycle idea, conducted the first attempts to prove the viability of open-cycle ocean thermal energy conversion (OC-OTEC). A preliminary test was done on land in 1928 in Belgium at Ougrée with a 60 kW gross power turbine. Warm-water flow came from a steel plant and the cold water source was the river Meuse.

In 1929, Claude undertook the building of a true OC-OTEC plant in Matanzas bay on the island of Cuba. Many of the thermal and electrical components were reused from the Ougrée experiment. Although faced with the challenging installation of a two-meter diameter cold-water pipe (which was only successful on the third attempt) the plant succeeded in producing 22 kW of gross power under a 14° C (25°F) temperature difference. But about ten days later, a storm broke the pipe and the project was abandoned.

The last enterprise by Georges Claude was in 1933 on the cargo ship **Tunisia**. The objective was to set up a floating ice-production plant offshore Brazil whose energy would have been generated by an OC-OTEC process. The plant included eight 275 kW turbines in line driving an 800-kW alternator and a 1200-kW ammonia compressor. Unfortunately the project was prematurely abandoned when the vertical cold-water pipe broke at an early stage of its installation [6].

In 1941, the French government resumed work on OC-OTEC and a state-owned company "Société Energie des Mers" (Ocean Energy Company) was established in 1948. The project dealt with a shore-based plant located in Abidjan, near the submarine canyon "Trou sans Fond" (bottomless hole). The plant was composed of two identical modules of 5-MW gross power each. A compact configuration was designed, in which the evaporator, the turbine, and the condenser were integrated in a single concrete vessel (Figure 6.2).

Many tests on the components proposed for the Abidjan plant were performed. Full-scale tests for installation at sea were conducted on actual lengths of cold-water pipe. Nisolle tested various types and geometries of evaporators while other tests were performed to assess performance of seawater predeaerators to be placed upstream of the heat exchangers [18].

In the 1950s, when the decision whether or not to build the plant had to be made after the development phase, the economic context did not favor OTEC and the project was stopped. A last appraisal of OC-OTEC was done in 1958 for a site in the French Antilles island of Guadeloupe, but with no positive result. Then the OTEC file remained closed until the 1970s when, following the 1973 oil crisis, OC-OTEC reassessment was undertaken in the United States and France.

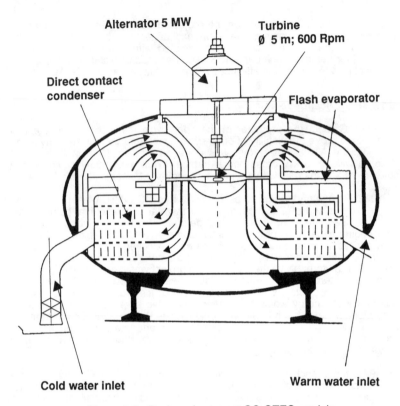

Figure 6.2 : Design of compact OC-OTEC module
for Abidjan power plant (1940s)

CNEXO the French state research organization in charge of ocean activities undertook feasibility studies of both closed- and open-cycle OTEC plants up to some tens of megawatt in 1978. Studies relating to the open-cycle plant were performed by various affiliated companies of CGE (French General Electric Company).

In 1982, IFREMER a newly created organism that had taken over CNEXO activities was given the mission of preparing the preliminary project file of an OTEC pilot plant to be located in the town of Papeete in Tahiti, French Polynesia. The net electrical power chosen for the plant was 5 MW, which was judged representative of the first expected market of OTEC plants ranging up to some tens of megawatts.

In the meantime, a number of French industrial companies that were interested in OTEC development had created a specialized joint venture, ERGOCEAN, which was selected to perform project studies on both closed-

and open-cycle options. In that context, Alsthom, an affiliated company of CGE, was given the mission of conducting the process and component studies for the OC plant option. The tasks included both design work on components and thermal tests in Alsthom's laboratories. The preliminary project file was made available in 1986. At that time, the energy context was not as much in favor of OTEC as it was in the 1970s, and the decision was taken not to proceed. Since then, all OTEC activity has been suspended in France, but some limited investigations into seawater desalination using the ocean thermal gradient were pursued until 1988.

U.S. federal government involvement in ocean energy research initially emphasized technology development for large, offshore, floating closed-cycle OTEC plants in the 100-1000 $MW_e$ range. Major testing of floating facilities at sea (discussed in Chapter 5) demonstrated the technical feasibility of the systems and provided extensive data on key components. With advancements in the OTEC technology base, analysis indicated that smaller-sized plants based on the open cycle offered an attractive potential ; therefore, the emphasis of the U.S. federal research program shifted toward open-cycle systems.

In 1979, the Westinghouse Corporation completed design studies for a 100-$MW_e$ OC-OTEC plant [21]. A design was developed for a floating system that used a large vertical-axis turbine, a channel flash evaporator, and a surface condenser, all enclosed in a reinforced concrete vacuum vessel. This design consisted of a cast-concrete containment structure that combined multiple structural functions with an axisymmetric layout for the major system components. The study concluded that an open-cycle plant could be a cost-effective alternative to the closed-cycle designs. It identified flash evaporation, direct-contact condensation, and use of innovative materials for the turbine and structural elements as ways to further reduce costs significantly. A study in 1985 explored the economic and technical viability of OC-OTEC for producing electricity or desalinated water or both, based on hardware reflecting the current state of the art [4]. It concluded that such plants could deliver desalinated water and electricity at a competitive cost to many island markets with plant sizes of less than 20 $MW_e$ and, in many cases, of sizes in the range of 2-10 $MW_e$.

Through the '980s. research on OC-OTEC components was conducted in a fresh-water dir. ~ contact heat- and mass-transfer laboratory at the Solar Energy Research Institute (SERI) to study and improve methods for transferring heat and mass with small driving potentials and at low operating pressures. A number of evaporator geometries were tested in this facility [2]. Extensive experiments were also carried out on various configurations of contacting media suitable for condensing steam at the low pressures applicable to OC-OTEC. The data from these experiments were used to validate detailed analytical models describing the performance of the direct-contact condensers [1]. The development of this analytical capability and the definition of hardware concepts that could achieve the performance potential inherent in the direct-contact processes were significant steps in development of OC-OTEC systems.

Since 1986, U.S. research efforts concentrated on scaling up the experimental apparatus to a thermal capacity of over 1 MW using seawater as the process fluid. This heat- and mass-transfer scoping test apparatus

(HMTSTA) is installed at the U.S. Department of Energy's Seacoast Test Facility (STF) on the grounds of the Natural Energy Laboratory of Hawaii. Tests in seawater [22] confirmed the heat- and mass- transfer performance of spout evaporators and structured packing condensers observed in fresh-water experiments and predicted by the analytical models [1] ; data on the performance of surface condensers and on seawater deaeration also were obtained in related experiments. Parallel activities were initiated in 1987 to develop a suitable OC-OTEC turbine, based on existing hardware and/or existing technology.

Efforts are currently focused on the design and construction of a net power-producing experiment (NPPE) to be installed at the STF in 1991-92. The NPPE will utilize the full seawater capacity of the STF.  The seawater systems at this facility can deliver volume flows of cold and warm seawater up to 0.41 m$^3$/s (6500 gpm) and 0.60 m$^3$/s (9500 gpm), respectively, to the OTEC equipment.   Another 0.43 m$^3$/s (6800 gpm) of cold seawater is delivered to other experiments, primarily in mariculture. Operation of a pre-commercial sized plant as is the NPPE would provide information on long-term reliability, availability, performance, and maintainability of the OC-OTEC system. Land-based and near-shore OTEC facilities in the size range of 1-15 MW$_e$ for production of electricity and potable water, are considered the most probable for early market penetration in insular regions. Long-term use of OTEC is envisioned to expand to floating, self-propelled plantships that would "graze" in the tropical regions and manufacture energy-intensive products such as hydrogen, ammonia, and methanol.

## 6.2.  Thermodynamic Processes and Recent Component Development

In open-cycle technology, the warm seawater is the working fluid. As is commonly observed when water boils at higher altitudes, the boiling temperature of water is a function of pressure. In the OC-OTEC cycle, warm surface seawater boils inside a vacuum chamber that is maintained at an appropriately low pressure. At the low OC-OTEC operating pressures, the density and energy content of the steam is very low; therefore, large flow passages and very efficient heat exchangers must be used and the turbine must be designed to efficiently extract the energy from the steam at the low operating pressures and convert it to electrical power.

Two open-cycle approaches have been investigated: (1) the Claude cycle, which is described in the next section, and (2) a more recent innovation commonly referred to as the mist-lift cycle. This process consists of flash evaporating the warm seawater at the bottom of a large evacuated lift tube. The mist, carrying seawater droplets, rises to the top of the tube, thus increasing the elevation (potential energy) of the water.  At the top of the tube, the mist is condensed back to a liquid, using the cold deep ocean water. To complete the cycle, electric power is generated when the condensed water flows down a tube, under gravitational pull, discharging through a standard hydraulic turbine. To date, only proof-of-concept and scoping tests have been conducted on the mist-lift cycle, and there is currently no activity on this concept.

## 6.2.1. Thermodynamics of the Open (Claude) Cycle

A block diagram of the open (Claude) cycle is shown in Figure 6.1 above. Warm surface seawater typically at 26°C (78°F) enters a vacuum chamber maintained at pressures around 2.1 kPa (0.4 psi) absolute. Because this pressure is below the vapor pressure of water at 26°C, part of the water boils in a flash-evaporative process. The remainder of the seawater, having been cooled by this evaporation, is discharged back to the ocean. The steam flows through a low-pressure turbine that is coupled to a generator, then to a condenser that operates at about 1.2 kPa (0.2 psi) absolute. The condenser uses cold seawater at about 6°C (43°F) pumped from depths of up to 1000 m (3300 ft) to condense the turbine exhaust steam. This condenser can be a direct-contact type, in which the seawater is mixed with the steam; or it can be a surface type, in which seawater is separated from the steam by an impervious surface.

Most thermal power conversion systems require vent systems to remove small quantities of noncondensable gas (mainly oxygen and nitrogen) from the working fluid. In OC-OTEC, the steam contains a significantly higher percentage of this gas because seawater contains small but significant quantities of gas in solution, from 15 to 20 milligrams per kilogram of water. When the seawater enters the low-pressure heat exchangers, a large portion of the gas comes out of solution and enters the vacuum vessel. The desorbed gas, along with air leaking into the vacuum chamber, accumulates in the condenser, blanketing the condensing surfaces and decreasing heat transfer. To maintain suitable operating pressures in the condenser, a sizable vent-compressor train is required to remove this gas.

In addition to these power system components, an OC-OTEC plant requires a vacuum containment structure to house the heat exchangers and, in some designs, the turbine. The open cycle also requires a seawater subsystem consisting of warm- and cold-water supply pipes, separate or mixed-water discharge pipes, and seawater pumps, as was discussed for the closed cycle in Chapter 5. Optional components of the OC-OTEC system include a mist-removal device between the evaporator and the turbine, to remove water droplets carried by the wet steam ; and two predeaerators, one in each of the cold- and warm-seawater inlet streams, located upstream of the heat exchangers. Predeaerators remove a portion of dissolved, noncondensable gas from the water stream at a pressure higher than that in the vacuum chamber. Because the water-side pressure loss through passive predeaerators can be negligible, the passive predeaerators can reduce vent-compressor power at no significant increase in parasitic pump power.

### 6.2.2 Component Development

The principal components of an OC-OTEC plant are shown in Figure 6.1 and described in the next sections. These sections summarize the operational requirements, the configurations considered, and the governing parameters that describe the performance of the components. All components except the turbine have been tested in fresh water and/or in seawater and their thermal and hydraulic performances are summarized in the respective sections.

The seawater systems are similar to those used for the closed cycle, as discussed in Chapter 5. The optional predeaerators require further investigation.

### 6.2.2.1. Flash Evaporator

The steam-production process in the evaporator occurs from the combined action of boiling and surface evaporation. Flash evaporation is usually quite violent as a result of the explosive growth of vapor bubbles from nucleation sites in the liquid; the bubble growth shatters the liquid continuum and yields a wide range of droplet sizes. Because of the irregular geometry of the interface, it is practically impossible to define or measure the surface area from which evaporation takes place. Therefore, heat transfer in flash evaporation cannot be described in terms of conventional heat-transfer coefficients. Instead, thermal effectiveness is used and is defined as the ratio of the temperature difference in the liquid stream to the maximum temperature difference that is thermodynamically available :

$$\varepsilon_{ev} = \frac{(T_{wi} - T_{wo})}{(T_{wi} - T_s^*)}$$

(6-1)
(see the list of notations
at the end of chapter)

The numerator is the temperature drop of the warm seawater as it flows through the evaporator ; the denominator is the driving potential or superheat, where $T_s^*$ is the temperature of seawater that is in equilibrium with the steam exiting the evaporator. Typically, this temperature is 0.31°C (0.56°F) higher than the fresh-water saturation temperature because of boiling point elevation, giving an ideal driving potential that is smaller in seawater than in fresh water. An effectiveness value of 1 implies that the steam leaving the evaporator is in equilibrium with the discharge seawater, and represents the theoretical maximum effectiveness.

A multitude of flash-evaporator configurations have been tested, from open-channel flow to falling films and jets. Results showed that high evaporation effectiveness can be achieved regardless of the inlet geometry, provided that the liquid is broken up to expose large interfacial areas and that the vapor/liquid interfaces are continuously renewed. Thus the design of an effective evaporator focused on the liquid distribution manifold to achieve low losses in liquid and gas pressure. Early work by Nisolle in 1947 and recent investigations at SERI in 1983 resulted in the selection of the "vertical spout" configuration as the preferred evaporator geometry for OC-OTEC systems. The spout evaporator consists of a series of vertical tubes distributed across the horizontal area of the evaporator chamber that carry the warm seawater into the chamber and expose it to the low pressure. Dramatic differences are visible in the spout flow without and with flashing. When the chamber pressure is high, there is no flashing, and the flow resembles a simple fountain flow. When the pressure is lowered and flashing occurs, bubbles begin to form as the seawater approaches the top of the spout. The water jet shatters into fragments and droplets, and steam escapes from the bursting bubbles. Most of the liquid transported upward falls back on the incoming liquid and the coherent liquid sheets are destroyed, becoming a spray of droplets. The vertical spout configuration results in easy vapor separation with minimal obstruction in the vapor path, low steam and seawater pressure drops, and the potential for a modular evaporator design that is not dependent on plant size.

A number of efforts to model analytically the heat and mass exchange in turbulent and shattering jets, falling sheets, circulating drops, or shattered sprays with mean droplet diameters have only been marginally successful in predicting the performance of the evaporator. The difficulties arise from a poor definition of the interfacial area and the periodic renewal of this surface, the extreme activity in the cap just above the spout, and the rapid decay of turbulence downstream. Fortunately, the vertical spout evaporator has consistently displayed very high thermal performance; therefore, accurate model predictions are not required for engineering applications.

Experiments have shown that the overall thermal performance of the vertical spout evaporator is primarily a function of the liquid and steam loadings, defined as the overall mass flow rates per unit planform area, as has been well documented for countercurrent cooling towers and other gas/liquid contacting devices. The effectiveness is not directly dependent on parameters such as spout velocity or diameter [13], but depends on the product of spout diameter squared and spout velocity, resulting from the liquid loading.

Figure 6.3 shows the dependence of thermal effectiveness on liquid loading for a one-spout and a three-spout configuration tested at the HMTSTA [23]. Results show that the effectiveness decreases with increasing liquid loading for both single and multiple spouts. For multiple spouts, spout-to-spout or spout-to-wall interactions decrease the overall thermal effectiveness from that obtained with a single spout for the same liquid loading. Data obtained by Nisolle [13] with up to 175 spouts and by Fournier [8] with three spouts are also shown in Figure 6.3. Nisolle tested spouts and also rectangular slots ; equivalent diameters are shown in the legend. Fournier observed qualitatively that the vapor could not readily escape when three spouts were present in the small test chamber, resulting in higher vapor pressure losses (lower thermal effectiveness) than for the single-spout tests. All multiple-spout geometries show very similar performance (within the respective error bars) regardless of the number, size, or spacing of the multiple spouts. The evaporator planform areas for the HMTSTA and for Nisolle's tests are similar ; therefore, the degree of spout-to-spout interaction among the 175 spouts of Nisolle is much larger than among the 3 spouts in the HMTSTA. The similar response at a particular liquid loading for all configurations tested implies that, once the fluid is brought into the chamber and it flashes, the inlet parameters are no longer important. It is the total quantity of water and the availability of free paths for the steam to escape that influence the thermal performance. This provides flexibility in the selection and geometrical arrangement of spout evaporators.

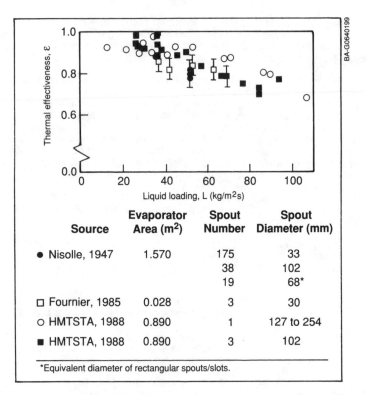

| Source | Evaporator Area ($m^2$) | Spout Number | Spout Diameter (mm) |
|---|---|---|---|
| ● Nisolle, 1947 | 1.570 | 175 | 33 |
| | | 38 | 102 |
| | | 19 | 68* |
| □ Fournier, 1985 | 0.028 | 3 | 30 |
| ○ HMTSTA, 1988 | 0.890 | 1 | 127 to 254 |
| ■ HMTSTA, 1988 | 0.890 | 3 | 102 |

*Equivalent diameter of rectangular spouts/slots.

Figure 6.3 : Influence of liquid loading on thermal
effectiveness of vertical-spout evaporator
(from Ref. 23)

The thermal effectiveness of a spout evaporator increases slightly with superheat and spout height. Superheats above 2°C (3.6°F) and a spout height of 0.5 m (1.6 ft) are required to achieve high evaporator effectiveness. Enhancement of the thermal effectiveness at the higher liquid loadings was obtained in fresh water by adding screens in the water path around the spouts [2]. A similar attempt during the HMTSTA seawater tests did not succeed because the foam produced in the seawater [8] curtails the effect of the enhancing screens.

In all experiments, hydraulic instabilities among spouts have never been observed. When viewed from above, the distribution of droplets from the 175 spouts tested by Nisolle is qualitatively indistinguishable from the field of droplets created by the three (much-larger-diameter) spouts tested at the HMTSTA. Because larger spouts can deliver the same quantity of fluid with lower hydraulic losses and less plumbing, this indicates a design preference toward the larger spouts.

Typical spout hydraulic losses are 0.7 m (2.3 ft), including spout height. Spout hydraulic loss coefficients (ratio of hydraulic pressure loss to the dynamic pressure of the flowing liquid) at the HMTSTA were about 20% higher than the single-phase flow predictions indicated. This increase appears to be caused by the two-phase bubbly flow at the top of the spout. Because of the foamy nature of the discharge seawater, the hydraulic loss coefficient for the discharge was 50% higher than the single-phase flow prediction. Discharge losses can be reduced by proper design of the piping for foamy flow.

Commercial chevron-type plastic mist eliminators mounted 2-3 m (6-10 ft) above the spouts provide suitable elimination of droplets for steam velocities through the mist eliminator of up to 19 m/s (62 ft/s). They result in small pressure loss and low droplet carryover. This is inferred from chemistry analysis of the condensate, which amply meets drinking-water quality standards [23].

### 6.2.2.2. Condenser

Two types of condensers can be used for the open cycle: direct-contact and surface condensers. The direct-contact condenser (DCC) is a gas/liquid contacting device in which steam exhausted from the turbine condenses directly on the cold water. These devices have gained recent prominence because they provide high volumetric heat-transfer coefficients at low cost [10]. In the surface condenser, the steam condenses on a metal surface that separates the steam from the cold seawater. Surface condensers have long been used in industry, and their performance is well characterized for those applications. Surface condensers cost more than direct-contact condensers for the same efficiency. However, by preventing the cold seawater from contaminating the steam condensate, they provide desalinated water as a valuable by-product to the plant's output of electric power. Direct-contact condensers can be used to produce desalinated water ; the steam must be condensed onto fresh water that is circulated in a closed loop, and the latent heat of condensation must be removed from this fresh water by cold seawater flowing through a separate liquid/liquid heat exchanger. For OC-OTEC plants producing only electricity, the DCC offers a large cost benefit. For fresh water coproduction, further research is required to determine a cost-effective arrangement.

The direct-contact condensation process is similar to what occurs in the OC-OTEC evaporator, except there is no shattering of the liquid to provide large interfacial areas for steam condensation. Therefore, research has concentrated on identifying geometric configurations that expose a large surface area of water to the steam (drops, rivulets, or thin sheets) while providing appropriate gas flow channels to keep gas pressure losses low. The condensation mechanism can be described as (a) a molecular mass transfer from the vapor to the liquid interface, with release of the latent heat of condensation ; and (b) simultaneous heat transfer from the interface, which is at an intermediate temperature, to the bulk of the liquid. The highest resistance to heat transfer can be from the liquid or gas phases. In the absence of noncondensable gases, the liquid-side heat-transfer resistance dominates. As noncondensable gas concentrations increase, the main resistance shifts to the gas side because the noncondensable gases are drawn to the interface by the condensing steam and they blanket the interface, retarding condensation. The thermal performance of the condenser can also be expressed in terms of its effectiveness, as was done for the evaporator. This is discussed below. In a direct-contact device the liquid flows by gravity ; the flow direction for the gas can either be cocurrent or countercurrent to the liquid. Countercurrent contacting devices are generally the most effective ; however cocurrent devices have an important application in OC-OTEC systems, primarily because of plant layout. Typically, the steam is exhausted from the turbine above the condenser, and large pipes direct the steam to the condenser entrance. If a single countercurrent condenser is used, these pipes must route the steam to the base of the condenser. Therefore, a cocurrent first stage in place of these pipes makes good use of this flow area and it can reduce the required planform area of the countercurrent condenser stage by 70-90 %. This approach has been implemented in the more recent U.S. designs.

Numerous geometries have been tested, from open trays and spray columns to falling thin sheets. Typically these geometries attain a condenser thermal effectiveness of up to about 0.6. This level of effectiveness is insufficient for the OC-OTEC process ; therefore, additional geometries have been developed.

In the IFREMER-ERGOCEAN Tahiti pilot plant project, the condenser work focused on the performance enhancement of falling jet condensers. The final solution was a geometry combining local crosscurrent flows of the steam through water jets in a global countercurrent pattern, as shown in Figure 6.4. The steam condensation takes place in modular structures consisting of a number of superposed perforated trays. The trays are arranged symmetrically with respect to the main longitudinal plane of the condensing structures. Cold water is delivered at the upper tray and then falls from tray to tray in the form of a curtain of water jets. The higher jets and trays are enclosed in a light cover. Steam is conveyed toward the lower trays, gets through the water jets, and then climbs upward to the highest tray in alternate crosscurrent flows. As steam condenses, its volume flow rate decreases. To maintain a high steam/water crossing velocity, which favors heat exchange, the number of superposed water falls that are crossed in parallel by the steam flow are gradually decreased.

Typically, the width of the trays is around 0.2 m (0.6 ft). Their length depends on the steam flow rate to be condensed in the modular structure, but it can reach several meters. The water jet diameter needs to be between 5 and 10 mm (0.2 and 0.4 in.). Along the width of a tray, steam encounters about 15 jets. Falling height of the jets is typically 0.2 m (0.7 ft). The modular structures are composed of 7 or 8 superposed series of jets.

This single-stage condenser design has been tested in the laboratory using a 30-kW thermal-power loop. This power level corresponds physically to a slice of one jet row in one of the symmetrical halves of the full-scale, modular condensing structure proposed for the Tahiti plant. The tests were performed both with fresh and salt-added water. They have shown a rather high level of thermal effectiveness of the condenser of up to 0.85. The temperature difference between the cold incoming water and the outflow of steam and noncondensable gases was about 1°C (1.8°F). About 98 % of the incoming steam flow was condensed, and total pressure drop was no more than 200 Pa (0.03 psi).

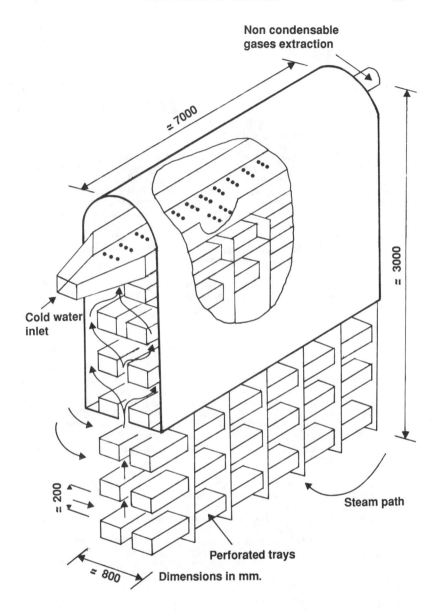

Figure 6.4 : Falling jets direct-contact condenser
design for Tahiti project (1985)

Random and structured packings have also been tested and have demonstrated thermal effectiveness up to 0.93. For the OC-OTEC system, structured packings offer the highest thermal performance with low gas-pressure losses [1]. These consist of a series of corrugated sheets made of metal, plastic, or gauze which form flow channels that are alternately oriented with respect to the vertical axis. The cold water is distributed over the top of the packing and flows along the sheets in thin layers ; surface renewal is provided by surface roughness and flow reorientation between channels. In contrast to the evaporator or to random packing, the interfacial area is fairly well defined by the geometry of the sheets.

Analytical modeling capabilities for direct-contact condensers have been developed only very recently ; industrial users have relied primarily on design guidelines obtained experimentally. The majority of the analytical work concentrated on spray or random-pack columns with single-component (heat or mass) transfer and at pressures well above ambient [10]. Detailed models for cocurrent and countercurrent flow with large latent heat transfer and at large concentrations of noncondensable gases were recently developed at SERI [1]. These models were verified extensively with fresh water and seawater at OC-OTEC operating conditions and accurately predict the DCC performance for engineering applications [23].

Two performance parameters are used to describe the direct-contact condenser : the thermal effectiveness and the vent ratio. The thermal effectiveness of the DCC is expressed as :

$$\varepsilon_C = \frac{(T_{wo} - T_{wi})}{(T_{si}^* - T_{wi})} \qquad (6\text{-}2)$$

where $T_{si}^*$ is the temperature of seawater in equilibrium with the steam incoming from the turbine. At condenser pressures, this is typically 0.28°C (0.50°F) higher than the fresh-water saturation temperature and gives an ideal driving potential that is larger in seawater than in fresh water. As is the case for the evaporator, an effectiveness of 1 represents ideal use of the available temperature potential.

Because the noncondensable gases are concentrated through the condenser, the steam can only be condensed to a value that is in equilibrium with these gases at the exhaust ; the remainder must be vented out with the noncondensable gases. The uncondensed steam is typically 50 % to 75 % of the exhausted gas mixture. The vent ratio V is the ratio of two total volumetric flow rates : the ideal (smallest) volumetric flow out of the condenser and the actual exhaust flow. It can be calculated by applying the ideal gas law to either component (steam or noncondensables) of the exhaust stream. Selecting the noncondensable gas for the derivation, the vent ratio can be expressed as the ratio of the actual noncondensable gas density to the ideal (largest) noncondensable-gas density. The density of this gas at the condenser outlet is $\rho_{io} = pp_{io} MW_i / RT_{io}$, where the partial pressure of the noncondensable gases is $pp_{io} = P_o - pp_{so}$. The total outlet pressure is $P_o$. The steam partial pressure is $pp_{so} = P_{sat}(T_{so})$ for saturated conditions. $MW_i$ is the molecular weight of the gases, R the gas constant, and $T_{io}$ the absolute outlet gas temperature. The ideal volumetric flow occurs when two conditions are met. First, the steam and noncondensable gases flowing through the condenser must incur no pressure losses, so that the outlet pressure $P_{o,ideal}$ equals the inlet pressure $P_{i,1}$ at the inlet of the first condenser stage. Second, the steam must be condensed to the thermodynamic limit, so that the steam exiting the second condenser stage is in equilibrium with the cold seawater temperature at that location. This implies that the outlet steam partial pressure is $pp_{so,ideal} = P_{sat} (T_{wi,2})$. Instead, the real condenser incurs pressure losses $\Sigma\Delta P$ and the steam exits at a saturation temperature that is higher than $T_{wi,2}$. Thus the ratio of total ideal volumetric flow to actual exhaust flow yields the following expression for the vent ratio in SI units :

$$ V = \frac{P_{i,1} - \Sigma\Delta P - P_{sat}(T_{so})}{P_{i,1} - P_{sat} (T_{wi,2})} \quad \frac{(T_{wi} + 273.15)}{(T_{so} + 273.15)} \qquad (6\text{-}3) $$

The vent ratio for OC-OTEC systems is very sensitive to pressure losses in the condenser because the overall available pressure difference is typically less than 500 Pa (0.07 psi). Ideal conditions are represented by $V = 1$, implying that all the steam that can be condensed has been removed from the exhaust flow.

The thermal and exhaust performance of the direct-contact condenser stages are complex functions of liquid and steam loading, noncondensable gas concentration, saturation temperature, stage geometry, and packing geometry. No simple design guidelines have yet been developed, as exist for surface heat exchangers and condensers. The reader is referred to the parametric studies conducted with the analytical models to review the performance of each stage in detail [1]. Here, appropriate operating conditions for OC-OTEC systems will be briefly summarized, based on the two-stage geometry and structured packings tested at the HMTSTA [23]. This plastic packing material is produced commercially and provides an effective surface area per unit volume of 98 $m^2/m^3$.

Figure 6.5 is a photograph of the HMTSTA condenser. The assembled countercurrent stage is at the center and it contains a height of 0.9 m (3 ft) of structured packing. Cold seawater is supplied centrally through the two white pipes near the top and it is discharged onto a perforated drip tray that distributes the seawater over the packing. Stand pipes in the drip tray allow the noncondensable gas and the small quantity of uncondensed steam to flow into the upper chamber and through the large white pipe at the top of the assembly so that they can be removed by the vacuum system. The cocurrent stage is coaxial with the countercurrent stage. A height of 0.6 m (2 ft) of the same type of packing material is being installed. The space below the two stages remains open to allow gas flow between the stages. This unit has a thermal capacity of 1.25 MW at OC-OTEC conditions.

At a typical steam loading around 0.4 $kg/m^2s$ (0.08 $lb/ft^2s$), the first (cocurrent) stage condenses 70 % to 90 % of the steam from the turbine with a constant thermal effectiveness between 0.9 and 0.95, depending on the concentration of noncondensable gas in the stream entering the condenser.

This concentration depends primarily on the gas liberation rate from the warm seawater in the evaporator. For a packing height of 0.6 m (2 ft), the pressure drop is less than 50 Pa ($7 \times 10^{-3}$ psi) and the steam saturation temperature at the outlet of the stage is typically 0.5°C (0.9°F) lower than at the inlet.

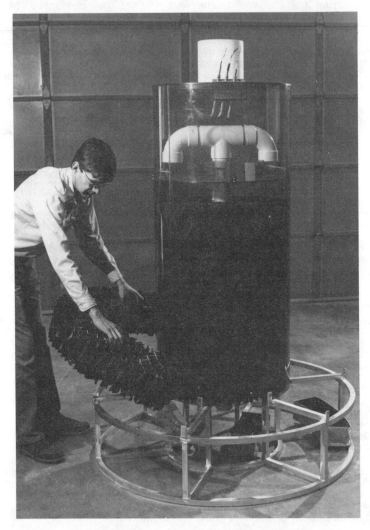

Figure 6.5 : Two-stage direct-contact condenser
assembly for a thermal capacity of 1.25 MW
at OC-OTEC conditions

The second (countercurrent) stage concentrates the noncondensable gases that evolve in the vessels and provides a close approach between outlet steam saturation temperature and inlet cold seawater temperature. For operation with seawater, this approach can readily be zero or even slightly negative. A small approach temperature is important to obtain high vent ratios. This stage typically condenses 95 % or more of the steam it receives from the first stage, over a broad range of liquid loading. The gas pressure drop is nearly a quadratic function of steam loading to this stage ; for steam loadings below 0.25 kg/m$^2$s (0.05 lb/ft$^2$s), the gas pressure drop is less than 50 Pa (7x10$^{-3}$ psi).

Figure 6.6 shows typical overall performance of a DCC in terms of vent ratio and overall thermal effectiveness. The data points represent operation over a broad range of steam and liquid loadings for the DCC, and they have been divided according to the inlet steam saturation temperature, which controls the turbine outlet temperature. An ideal condenser would operate at the upper right corner of the graph, where both performance parameters have a value of one. A real condenser can operate within 10 % of this thermodynamic limit.

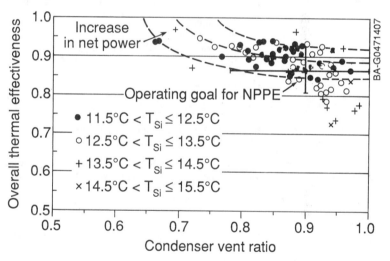

Figure 6.6 : Thermal and venting performance
of direct-contact condenser
with structured packings (from Ref. 23)

An increasing value of V represents a reduction in steam exhausted with the noncondensable gases, implying lower exhaust power consumption. An increasing value of thermal effectiveness represents a reduction of cold seawater use, implying lower seawater pumping power. Exhaust and seawater pumping power account for the majority of parasitic power losses in the OC-OTEC system. Therefore, the minimum system parasitic losses occur when the condenser operates at the upper right corner (the ideal performance limit). Constant system parasitic losses can be represented by a set of curves that encircle this corner, as shown by the dashed lines in Figure 6.6. The exact shape of the curves depends on the specific performance of other system components, but they tend to be oval as shown. Curves that are further away from the upper right corner imply larger overall parasitic losses, thus less net-power output from the system. By adjusting geometrical and operating parameters of the condenser stages, the condenser can be designed to operate at any point along one of these parasitic curves, which would result in the same system parasitic losses. Thus system analysis must give the appropriate selection of condenser operating conditions for a specific system and application.

Typically, more than 99 % of the steam is condensed in the two-stage, structured-packing condenser, and the overall gas pressure drop can be maintained below 100 Pa (0.015 psi) for packing heights of less than 1 m (3 ft). Total cold-water hydraulic losses of less than 2 m (6 ft) can be obtained using vertical spout distributors on the cocurrent stage and drip trays on the countercurrent stage.

The second condenser option is to use a surface condenser, in which the cold seawater is physically separated from the condensing steam by an impervious surface. Because the cost of a surface condenser is much higher than the cost of a direct-contact condenser, it is generally not considered for OC-OTEC plants producing only electricity. However, surface condensers become more attractive when desalinated water is a valuable by-product. The cost of the condenser is affected by two items : the thermal performance and the cost of the material. Therefore, research has concentrated on improving the performance by suitable design and on identifying low-cost materials for operation with seawater. The research conducted for closed-cycle OTEC heat exchangers on both these topics is directly applicable here, and the reader is referred to Chapter 5 for a detailed discussion; only those characteristics that are specific to the open cycle will be discussed here.

Surface condensers have been used extensively in industrial power plants, and their performance is well characterized for those applications, especially for shell-and-tube condensers. However, OC-OTEC operating conditions are outside the range of applicability of the computer models and the design guidelines established for power-plant condensers, primarily because the concentration of noncondensable gas is much higher than that encountered in conventional applications. Blanketing of the condensing surfaces, nonuniform spatial distribution of gases, and formation of stagnant pockets of noncondensable gas significantly decrease the performance of the condenser and increase the gas pressure losses.

As is the case for the DCC, the surface condenser for OC-OTEC applications is a condenser with a well-defined steam path where the noncondensable gases are swept to the condenser outlet while maintaining low gas pressure losses. The coldest seawater must be available where the noncondensable concentration is highest, in order to ensure sufficiently high condensation rates for the condenser. Plate-fin or channel-fin designs with multiple passes and cross flow meet these requirements and have been proposed for OC-OTEC systems and investigated by the Argonne National Laboratory [14]. The number of fins per unit width of passage increases along the steam flow path to compensate for the decrease in the rate of condensation per unit area and the decrease in the steam velocity along this length. This geometry has not yet been tested at OC-OTEC conditions. Two other geometries have been tested : a cross-flow condenser with dimpled parallel plates, and an extruded-channel condenser with fins, operating with refrigerant R-12. Tests were inconclusive in quantifying the effect of noncondensable gases on the performance of these surface condensers at high concentrations of noncondensable gas [23]. At lower concentrations, the dimple-plate condenser performed as predicted by a conventional model for this unenhanced, cross-flow geometry.

### 6.2.2.3. Turbine

Turbines for OC-OTEC systems represent a substantial fraction of the total cost of the system, because the low density and low energy content of the steam make their size proportionally large. OTEC thermodynamic boundary conditions indicate that the turbine does not undergo significant thermal stresses or fluid-mechanical loading. For a fixed available steam temperature difference, the turbine power is determined entirely by the turbine's mass flow and efficiency, as described below. Availability of high steam mass flow from a system's perspective is discussed in Section 6.2.3.

Conventional turbine geometries used in industry, such as axial, radial, and mixed-flow configurations, have been considered for OC-OTEC. All three blade types can be developed into viable options. For low-pressure and high-mass-flow operation, the axial design is commonly applied in which low-pressure stages with high-efficiency blades are used. Axial designs have received the most attention from industry because staging is easy to arrange up to the number of stages required by the application. However, the OC-OTEC conditions can be satisfied by a single stage because of the low available pressure ratio. Westinghouse proposed an axial machine for the first OC-OTEC turbine [21]. This 140-MW$_e$ machine, with a diameter of 44.5 m (146 ft) and blades 12.2 m (40 ft) long, is outside the realm of the current state of the art. In fact, the blades are about 10 times longer than the largest existing blades.

Recent studies in France and the United States have suggested that the low-pressure stages of a nuclear turbine could be adapted, through appropriate redesign of the stator row, to operate satisfactorily as an OC-OTEC turbine for systems up to a few megawatts. Because such a turbine would be well characterized thermally, structurally, and dynamically, it is considered an option for a prototype open-cycle power plant. This option was selected for the IFREMER-ERGOCEAN pilot plant project in Tahiti (see Section 6.3.1). Alsthom-Rateau reviewed and assessed the low-pressure turbine blades they had previously designed in order to select which of them would suit typical steam conditions of OC-OTEC.

The steam temperature (or absolute pressure) at the turbine inlet and the temperature (or pressure) drop through the turbine provide the two steam properties that determine the dimensions of the turbine : first, the isentropic energy, which is the ideal energy that could be recovered from a unit mass of steam ; and second, the specific volume of steam at the turbine outlet. Isentropic energy is related to steam injection velocity at the outlet of the turbine fixed nozzles. To reach an optimum efficiency for axial flow, the injection velocity should be about twice the peripheral velocity of the turbine wheel at the blade root. Given the turbine revolution speed, which equals the electrical-grid frequency or half that value, the wheel diameter at the blade root is directly related to the specific energy of the steam. Steam mass flow rate is deduced from the gross power to be reached assuming a first estimate of turbine global efficiency. The volumetric flow rate and the steam velocity at the wheel blade outlet determine the required area of the turbine wheel's outlet section. This area requirement is combined with the base diameter to calculate the required blade height. For typical OC-OTEC conditions, the maximum turbine inlet temperature expected is 25°C and the maximum temperature drop is around 15°C, giving an upper bound for the isentropic energy of steam around 130 kJ/kg, while the specific volume at the turbine outlet can reach a value as high as 110 m$^3$/kg. Based on these conditions, the assessment of available blades led to the conclusion that a discrete series of turbine wheels would suit OC-OTEC conditions with global turbo-alternator efficiency reaching up to 0.8. To reach higher unit power, the turbogenerator could be composed of two turbine wheels positioned at the end of the alternator shaft (see Figure 6.7).

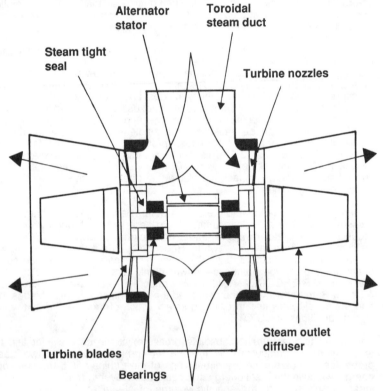

Figure 6.7 : OC-OTEC turbogenerator
(GEC-Althom Rateau design)

For a rotation speed of 3600 rpm, the wheel-base diameter would be about 1.2 m (3.9 ft). A discrete series of turbogenerators of the design described above could deliver gross power from 1 to about 2 MW, with blade heights ranging from 0.6 to 0.9 m (2 to 3 ft). Higher gross power could be obtained at the rotation speed of 1800 rpm. A discrete series of turbogenerators could be made available from 2 to around 4.5 $MW_e$ with blade heights ranging from 0.7 to 1.4 m (2.3-4.6 ft). Similar units would be available for rotation speeds of 1500 to 3000 rpm corresponding to a 50-Hz electrical frequency, as used in Europe. Of course, turbine performance will vary around the typical values shown above, depending on the exact values of steam conditions. A detailed assessment should be carried out for each specific case.

For larger OC-OTEC systems, the size and applicability of the low-pressure stage of conventional turbines have reached the limits of practicality because of the tooling costs associated with larger blades ; therefore, new designs must be sought. Several innovative design configurations or construction concepts have been analyzed by three university study teams, using all three possible turbine geometries, to investigate potential reductions in turbine costs and efficient integration of the turbine into the system [19], as described below.

Building on the Westinghouse design [21], one team considered a single-flow, single-stage, vertical axis turbine built of high-strength, lightweight composite materials such as fiber-reinforced plastics with urethane foam. Fiber-reinforced plastics and composites have been used successfully in military helicopter applications in blades approximately 12 m (39 ft) long. Mild OTEC steam conditions also permit a simplification of the highly expensive, complex, high-pressure housing usually associated with steam turbines. Westinghouse proposed using a special formulation of commercially available, high-density reinforced concrete. This concrete has relatively low air infiltration rates and can be cast to a very smooth surface finish without polishing.

A second team analyzed mixed-flow turbine blading with radial inflow of steam and exhaust along the axis of rotation. This type of blading is most often seen in gas compressors rather than in turbines. A notable exception is the turbocompressor found in many automotive vehicles. It is not used in conventional steam plants because of the complexity in staging the units for the pressure ratios used in these plants. One advantage of the radial-inflow concept is that the specific mechanical energy of the stage is higher than that of the axial stage, which means a smaller-sized turbine can be used and the radial rotor is inherently stiffer. This circumstance permits the blading to be fabricated from either thin, rolled sheet metal or cast plastic, which would reduce the investment cost considerably. This geometry was investigated further and it was selected for the NPPE system in Hawaii (see Section 6.3.2).

Cantilevered-blade designs were also examined, with radial inflow and radial outflow (squirrel cage), as well as cross flow. All configurations use a simple two-dimensional blade. The uniform cross section along the blade length can be designed to be the same for rotors and stators (nozzles) and can be fabricated inexpensively from low-cost materials. The crossflow turbine is essentially a two-stage, partial-admission, radial-flow turbine. Each stage uses rotors in a semicircular arrangement. The first stage acts as a radial-inflow design and the second stage acts as a radial-outflow design. The unique aspect of the crossflow configuration is that both stages employ the same rotor blades, which are mounted around the circumference of the rotor disk. The connection between the stages is nothing more than flow across the cylindrical space formed by the rotor blades.

One of the important conclusions drawn from the innovative-design study was that maximum efficiency is not the overriding criterion for choosing the design and operating parameters for an open-cycle OTEC turbine (in contrast to conventional power plants) when the use of existing hardware is considered. For example, the results showed that an axial turbine operating with a 33 % higher steam throughput and a 7 % lower efficiency than the most efficient configuration provides a cost-effective option for an open-cycle OTEC system.

### 6.2.2.4. Vent System

The vent system must continuously remove noncondensable gases and residual steam from the condenser and exhaust them to the atmosphere. Noncondensable gases are present in the incoming warm and cold seawater ; at the subatmospheric heat exchanger pressures, the dissolved gases are supersaturated, and virtually all of the noncondensable gases come out of solution. In addition, a small amount of air leaks into the vacuum enclosure from the atmosphere.

The vent system consumes a large fraction of the parasitic power in an OC-OTEC plant. Therefore, efficient components and drives are required. Two methods of gas compression have been proposed for OC-OTEC systems. The first is a multiple-stage mechanical compression train. The second uses discharge seawater to hydraulically compress the gas after an initial mechanical compression stage. The hydraulic compression option has potential for reducing the overall power consumption but requires implementing a large research and development effort.

In the multiple-stage mechanical compression train, interstage gas coolers are important in minimizing the power consumption. Cooling the gas before it enters into the next compression stage minimizes the work required to compress a given mass of gas. In addition, the intercoolers can also be used to condense some of the residual steam, thereby reducing the gas mass flow rate that must be processed by the downstream compression stages.

Positive-displacement compressors are typically used at vacuum levels compatible with OC-OTEC, but applications are not overly concerned with efficiency. Rotary lobe and rotary vane machines, with liquid ring pumps acting as the final stage, have efficiencies of 25 % to 36 %. The primary design advantages of existing positive-displacement compressor hardware are reliability and flexibility of operation. Commercial positive-displacement units typically have only two or three stages, resulting in stage pressure ratios that are higher than desirable for OC-OTEC applications.

Commercial centrifugal compressors are not designed to operate at the low pressures required for OC-OTEC, and because they are designed for gases of higher density, the drive systems and bearings are sized for large power requirements. The mechanical losses associated with existing centrifugal-compressor designs are excessive for low-density gas applications.

Centrifugal compressors with radial-blade rotors, which are normally used in a number of industrial processes to recompress steam, can provide an energy-saving solution. These machines (e.g., Alsthom-Rateau) are manufactured in a discrete series of radial-wheel diameters ranging from 0.2 m to about 1 m (0.6 to 3.3 ft). Compression ratios can reach values from 1.5 to 2.5 with volumetric flow rate reaching up to around 30 $m^3$/s (63,500 acfm) for the biggest machine. Therefore, about six compression stages are necessary in an OC-OTEC noncondensable venting system to compress the gases from about 1 kPa to atmospheric pressure. These machines are run at high rotation speeds reaching 18,000 rpm for the smaller ones. A multiplier gear train is necessary to drive the compressors with electrical motors rotating at 1800 or 3600 rpm. Mechanical losses are encountered in these gear trains. Transmission efficiency is around 0.7, and polytropic efficiency of radial compressors reaches values between 0.7 and 0.75.

### 6.2.3. System Power Output

Turbine gross power output is proportional to the product of steam flow and its temperature drop across the turbine ; therefore, there is a specific steam flow that maximizes the power output. Figure 6.8 shows how the overall resource temperature difference $\Delta T_r$ can be apportioned to the heat exchangers and to the turbine, and how this affects the turbine power output. This example assumes an overall $\Delta T_r$ of 20°C, between inlet warm seawater at 26°C and inlet cold seawater at 6°C, typical of the Hawaii site. Other assumed operating conditions are shown in the legend. The abscissa represents the steam flow that can be produced at a fixed seawater flow rate. The quantity of steam produced is on the order of 0.5 % of the warm seawater flow. The ordinate represents steam and seawater temperatures as well as gross power output for a fixed-geometry turbine. $\Delta T_{ww}$ is the temperature drop incurred in the warm seawater as it flows through the evaporator to produce the steam. $\Delta T_{cw}$ is the cold seawater temperature rise through the condenser. $\Delta T_{turb}$ is the steam temperature difference available for the turbine. As more steam is produced for the same seawater flows, $\Delta T_{ww}$ and $\Delta T_{cw}$ increase, thus reducing the temperature difference available for the turbine. The small differences $\delta_{ev}$ and $\delta_c$ between each pair of straight lines (differences between seawater outlet temperature and steam temperature) represent the inefficiencies in the heat exchangers. These reduce the usable overall temperature difference to $\Delta T_u = \Delta T_r - \delta_{ev} - \delta_c$. The high thermal performance of the OC-OTEC heat exchangers gives $\Delta T_u \sim 0.93 \Delta T_r$ as can be seen in Figure 6.8. Because the system operates with seawater, the boiling-point-elevation effect is evident at zero steam flow rates.

The power output at the generator is $P = \eta_{TG} Q_s \Delta h_{is}$. Assuming a constant turbine efficiency and constant thermal effectiveness (Eqs. 6-1 and 6-2) for the heat exchangers, the maximum power output can be calculated, using the relation $\Delta T_{cw} = \Delta T_{ww} Q_{ww}/Q_{cw}$, and can be expressed as

$$P \propto Q_{ww} \Delta T_r^2 \qquad (6-4)$$

where $Q_{ww}$ is the mass flow rate of the inlet warm seawater and $\Delta T_r$ is the overall temperature difference of the resource, between incoming warm- and cold-seawater streams. The proportionality factor in Eq. 6-4 is a function of turbine parameters, heat exchanger effectiveness, and the ratio of warm to cold seawater flow rates ; for the OC-OTEC system currently envisioned, it is on the order of 1 $W_e(kg/s)^{-1}(°C)^{-2}$.

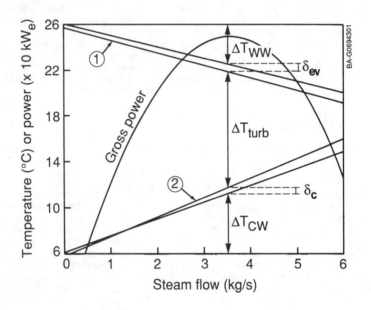

Steam temperature at
(1)  inlet of turbine
(2)  outlet of turbine
Warm seawater flow: 620 kg/s
Cold seawater flow: 420 kg/s

Evaporator effectiveness: 0.90
Condenser effectiveness: 0.86
Turbine efficiency: 0.85

Figure 6.8 : Variation of gross power for constant resource temperatures and seawater flows

The general relation for power output can be expressed in terms of the other OC-OTEC process variables as :

$$P = \frac{R_Q}{R_Q+1} \; Q_{cw} \, R_T \, (1 - R_T) \, \Delta Tu^2 \; \frac{C_p}{273 + T^*_{s,ev}} \; \eta_{TG}. \qquad (6\text{-}5)$$

with the following definitions:

. $R_Q = Q_{ww}/Q_{cw}$ the ratio of warm to cold seawater mass flow rates.

. $R_T = \Delta T_{turb}/\Delta T_u$ the fraction of the useful process temperature difference that is allocated to the turbine.

. $\Delta T_u = \Delta T_r - \delta_{ev} - \delta_c = \Delta T_{ww} + \Delta T_{turb} + \Delta T_{cw}$ the useful process temperature difference.

. $\Delta h_{is} = \Delta T_{turb} \, h_{fg} /(273 + T^*_{s,ev})$ the isentropic enthalpy drop across the turbine (valid for small $\Delta T_{turb}$ in the OC-OTEC range).

Equation 6-5 shows that for a given value of $Q_{cw}$, the gross power is a maximum when $R_T = 0.5$ (half the useful resource is allocated to the turbine, half to the heat exchangers), and increases with $R_Q$ (increasing warm-water flow). Because cold seawater requires more pumping power, therefore it is more expensive, $\Delta T_{cw}$ is always larger than $\Delta T_{ww}$ by a factor of nearly two. System analysis must guide this selection.

Part of the gross power produced by the generator is used to run the seawater pumps and the vent compressor. The remainder is the net power output of the plant. For large multimegawatt-sized plants, net power can be projected to be about 70 % to 75 % of the gross turbine output. Because warm seawater temperature cycles slightly during the year, it is not possible to produce the rated power during the entire year. This variability and the scheduled shutdown for plant maintenance result in a plant capacity factor of 0.85, which is very high for a solar plant.

### 6.2.4. System Models

A thermoeconomic study of OC-OTEC plants producing both electricity and desalinated water was conducted to determine the minimum plant size required to offer a viable option in six potential sites [4]. Figure 6.9 shows the resulting fractional component cost for an OC-OTEC plant optimized to produce electricity, using direct-contact heat exchangers. Some of the fractional costs, mainly for the seawater systems, differ from those found in other studies (e.g. see Chapter 7). The economics depend strongly on the assumptions made for plant design and for the component costs that are estimated from existing equipment [4]. Because no large OC-OTEC demonstration plant has yet been built, this gives a measure of the uncertainties in the various cost estimates.

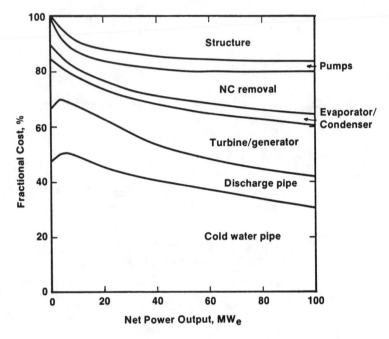

Figure 6.9 : Fractional component cost
of single-stage direct-contact condenser
plants optimized for power output (from ref. 4)

A comprehensive computer model exists [15] to describe the performance and projected costs of an OC-OTEC system. The model was developed at SERI for desktop microcomputer applications and written in FORTRAN. It consists of subroutines that contain the performance modeling equations for each of the components depicted in Figure 6.1. At this time, it does not contain a module for a surface condenser producing potable water. The subroutines and their related data bases are updated periodically as new experimental information is obtained. The system-level analysis can perform single-point runs, parametric sweeps, or optimizations. The primary criterion for the optimization is capital cost per unit power output and the optimization is carried out using a multivariant, step-wise, iterative routine that considers 14 parameters.

The cost algorithms are based primarily on the parametric studies of Block and Valenzuela [4] as a function of equipment size, with some modification for those components or portions thereof that were not studied by them.

Costs such as engineering, land, support buildings, and controls are estimated separately, based on the application. The output of the model allows a designer to identify the preferred operating conditions for each component, and to determine the sensitivity of plant cost to the projected cost and performance of these components.

A similar thermoeconomic model has been developed and used by IFREMER-ERGOCEAN to balance and optimize the characteristics of the OTEC Tahiti project [20].

### 6.3.   Recent OC-OTEC Plant Projects

6.3.1. ERGOCEAN/IFREMER 5-MWe Tahiti Project

The Tahiti project was initiated in 1983. The objective was to build an OTEC pilot plant powerful enough to be representative of plants having electricity-generating capacities up to some tens of megawatts, which were foreseen to be the first segment of the OTEC market.

Plant size was chosen for 5-MW$_e$ net output [9]. In fact, two options were considered during the project studies, one for closed and one for open cycle ; the cycle was to be chosen at the end of the predesign phase. The power goal was to be obtained for the maximum seasonal temperature difference at the Tahiti site, which is 24°C (75°F) when warm water reaches 28°C (82°F) and cold water is pumped from a 1000-m (3300-ft) depth.

Because of turbine size limitations, the OC-OTEC plant design is composed of two identical and autonomous modules (see Figure 6.10). Each module is essentially composed of a large vacuum vessel divided into three compartments, which contain the evaporator at the center and two condensers at the sides. Warm, cold and discharge waters are delivered through open-air channels excavated in the ground, and power components are positioned on a frame structure overhanging the channels. Seawater is pumped into the vacuum vessels and is evacuated by the barometric effect. The turbogenerator unit of each module is above the vacuum vessel ; it is of the type described in Section 6.2.2.3, with two turbine wheels at the ends of the alternator shaft. Steam leaving the evaporator is delivered to the turbine by a toroidal channel, and the two low-pressure steam flows are conveyed to the roof of the condenser compartments through diffusers and right-angle elbows.

The gross power of each turbogenerator unit is 3.3 MW. This power is generated by a 40-kg/s (88-lb/s) steam flow under a 13°C (23.4°F) temperature difference. Steam flow is directed to two turbine wheels with an outside diameter of nearly 4.5 m (14.8 ft) and blade height of 1.1 m (3.6 ft). The steel vacuum vessel is roughly 35 m (115 ft) long, 13 m (43 ft) wide, and 3.5 m (11 ft) high.

The floor area of the evaporator is about 200 m$^2$ (2150 ft$^2$). It is fitted with 22,000 spout tubes of 30-mm (1.2-in.) diameter delivering 7.5 m$^3$/s (118,900 gpm) of warm water. Each condenser is supplied with 2m$^3$/s (31,700 gpm) of cold seawater which is delivered to seven condensing structures with superposed perforated trays of the type described in Section 6.2.2.2.

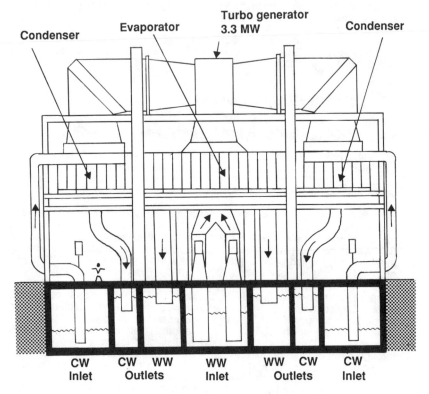

Figure 6.10 : Design for Tahiti project
3.3 $MW_e$ gross power

The mass flow rate of noncondensable gas to be extracted from the two condensers is about 0.2 kg/s (0.4 lb/s). Because the mixture to be vented out is composed of twice as much steam as noncondensable gas, the volumetric flow rate reaches 54 m³/s (114,440 acfm). The gas flow is compressed to atmospheric pressure through a venting system composed of six compressor stages with a number of intermediate condensers/coolers.

The power balance of the module shows that around 10 % of the gross power is consumed for venting, 10 % for pumping cold water, and 4 % for pumping warm water, so that the net available electrical power is roughly 75 % of gross power.

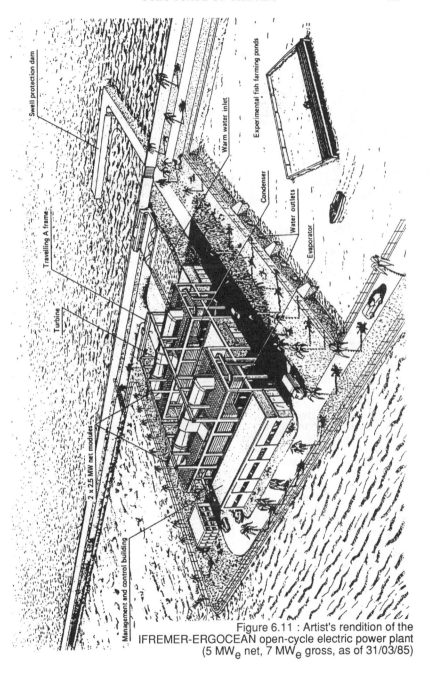

Figure 6.11 : Artist's rendition of the IFREMER-ERGOCEAN open-cycle electric power plant (5 MW$_e$ net, 7 MW$_e$ gross, as of 31/03/85)

An optimization study conducted at the end of the design phase showed that globally it would be more economical to pump cold water from only a 700-m depth, even if its temperature were 1.5°C higher than at the 1000-m depth. The reason is that the cold-water pipe length is significantly reduced. The saving on the cold-water pipe investment cost overrides the increase of the power subsystem cost, which results from the necessarily larger dimensions because of the reduced overall temperature difference, which also results in a loss of some points of global efficiency. This description does not take into account the small adjustments of power-component characteristics that are required because of the new operating temperature and flow values of the optimized process.

Figure 6.11 shows an artist's conception of the projected plant. As stated in Section 6.1, the preliminary design was completed in 1986 but the decision to proceed into detailed studies and plant construction has been postponed.


### 6.3.2. Net Power-Producing Experiment

The U.S. federal research effort is currently focused on the design and construction of a plant to investigate heat- and mass-transfer processes in large heat exchangers and to assess the viability of OC-OTEC systems [3]. The Net Power-Producing Experiment (NPPE), a cooperative effort between SERI and the Pacific International Center for High Technology Research (PICHTR), will be conducted in two stages. First, the heat exchangers will be tested to verify their performance and to determine the interaction among the system components. Then a turbine will be installed and the system's power-generating capability will be tested. The goal of the NPPE is to produce a minimum of 40 kW of net power. Construction of the NPPE has begun in the last quarter of 1991 for a completion in the second half of 1992. Figure 6.12 shows the general arrangement of the NPPE.

The NPPE system is designed around a single, vertical-axis, mixed-flow turbine (radial inflow, axial outflow). The rotor is 2.9 m (9.5 ft) in diameter. The rotor is supported by a vacuum vessel 7.2 m (24 ft) in diameter and 7.62 m (25 ft) high, which is the structural backbone of the system. The electrical generator is an air-cooled, salient-pole, synchronous generator located above the turbine assembly and outside the vacuum enclosure.

Steam is produced in an annular flash evaporator at the periphery of the vacuum vessel, which consists of 12 vertical spouts, each 0.24 m (10 in.) in diameter. The steam flows up from the evaporator, through a mist eliminator located at the turbine inlet flange, and enters the turbine radially inward. The steam exits the turbine axially in the center of the vessel. A conical exhaust diffuser with center body is used to recover most of the kinetic energy of the steam discharged from the turbine. The diffuser is 3.65 m (12 ft) long and provides an annular entry to the first condenser stage.

The condenser is a direct-contact, structured-packing condenser composed of two coaxial stages. The first stage fills the outer annular space of the condenser assembly, between the diameters of 4.11 m (13.5 ft) and 2.43 m (8 ft), and it receives steam from the turbine diffuser. It operates in cocurrent flow, with seawater and steam flowing vertically downward through 0.61-m-(2-ft)-high packing material. The second stage is in the center of the condenser assembly and operates in countercurrent flow to condense the steam left over from the first stage, so that less pumping is required for the vacuum vent system. The countercurrent stage contains the same packing material as the first stage. It is smaller in volume but longer, with an overall packing height of 0.91 m (3 ft). Cold seawater is supplied from below the condenser through a pipe at the center of the structure that rises above the condensers and that delivers all the seawater. The distribution manifold provides cold seawater to both stages through a drip tray for the countercurrent stage and a series of radial pipes for the cocurrent stage (not shown in Figure 6.11).

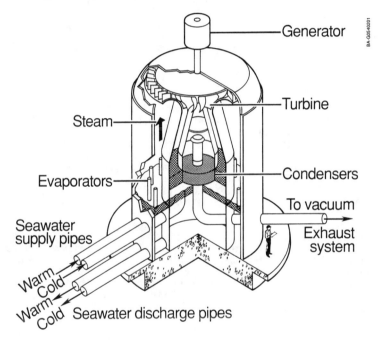

Figure 6.12 : General arrangement of components
for the Net Power-Producing Experiment
(from Ref. 3)

The noncondensable gases liberated from the seawater streams and a small amount of uncondensed steam are compressed and exhausted using a vacuum vent system. A five-stage compression system is planned. The first four stages are centrifugal compressors custom-designed for the NPPE to limit power consumption ; the last stage is a commercial positive-displacement compressor. Variable-frequency drives will be used to adjust the operating conditions of the stages and provide the flexibility to vary the exhaust flow rates required for the proposed heat exchangers and system operation tests. Direct-contact and surface intercoolers will be used between compression stages. A small surface condenser will be used outside the vacuum enclosure to conduct tests on desalinated-water production.

Figure 6.8 in Section 6.2.3 shows the nominal operating conditions for the NPPE. System analysis projects a gross power of 213 kW and a net power of 84 kW for the system at the baseline design conditions. Because the NPPE is small for an OC-OTEC system, the parasitic losses are a much higher fraction of the gross power than the 25 % to 30 % projected for a multi-megawatt system. Major design points for the NPPE are : 0.90 for evaporator effectiveness, 0.86 for condenser effectiveness, 0.85 for turbine total-to-total efficiency, 0.92 for generator efficiency, and 0.60 for exhaust system efficiency.

## 6.4.  PRODUCTION OF FRESH WATER

### 6.4.1. Fresh Water Coproduction and Single-Stage Desalination Unit

As stated in Section 6.2.2.2, the direct-contact condenser of an OC-OTEC generating plant can be replaced partly or totally by a surface condenser. Steam condensate thus provides a source of fresh water as a by-product of electricity generation. In the case of a plant with only a surface condenser, the ratio of fresh-water to electricity production capacity depends on site conditions and cycle definition ; typically this ratio is between 1200 and 1500 m$^3$/day (0.3 and 0.4 mgd) per each megawatt of net electrical capacity that is, 50 to 60 m$^3$ (13,210 to 15,850 gal) of fresh water can be produced for each megawatt hour delivered to the grid.

As described in Section 6.2.3, the steam temperature difference at the turbine is about half that between incoming warm- and cold-water temperatures. Thus the two seawater streams exiting the plant still maintains about half the initial temperature difference. The remaining temperature difference could be used in a simple single-stage evaporating/condensing apparatus, which would roughly double the electrical plant fresh-water production. Downstream of the desalination module, water flows would be nearly at the same temperature.

### 6.4.2. Multistage Desalination Units

Multistage processes can be used to more extensively exploit the thermal potential resulting from differences in warm- and cold-seawater temperatures. When large water flow rates are already available, as is the case downstream of an OTEC power-generating plant, two processes have been proposed. Initially multistage systems were designed for desalination using the warmed water flowing out of the condensers of thermal power plants; they can also be used in the OTEC context.

The low temperature multistage flash process [17] uses two equal flows of warm and cold water in a counterflow pattern (see Figure 6.13). Warm water flows through a series of vacuum chambers where flash evaporation takes place at gradually decreasing temperatures. Cold water flows through a series of condensers located in the upper region of each vacuum chamber.

The ratio of either the warm- or the cold-seawater flow to the fresh water production is about 70 for an available temperature difference of 15°C. Dissolved gases evolve from the warm water under vacuum, and their recompression to atmospheric pressure requires a rather high energy consumption. To reduce that consumption, a predeaerator, operating at a pressure around 10 kPa, is installed upstream of the evaporator. When this process is used in connection with a conventional thermal power plant, noncondensable compression is made by several stages of steam ejectors because steam is readily available. For OC-OTEC operation, compression should be done by electrically-driven mechanical compressors instead of steam ejectors. The specific electrical consumption is rather high because of the large amount of warm-seawater outgassing in the process.

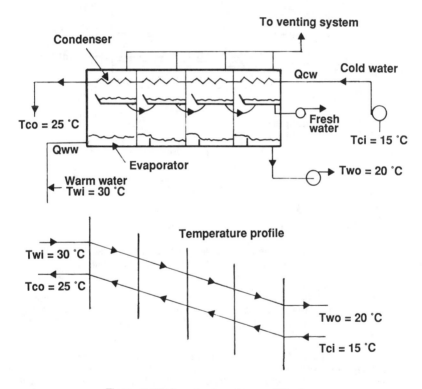

Figure 6.13 : Low-temperature multiflash
desalination process

The horizontal sprayed tubes process has been designed [12] to reduce the amount of outgassing from the warm water. A quantity of water equivalent to only about twice the fresh-water production is sprayed in the evaporation chambers onto a tube bundle which carries the warm water (see Figure 6.14).

The volume of desorbed air is thus considerably reduced when compared to the multiflash process. Specific electrical consumption of the sprayed-tubes process is about 4 kWh/m$^3$, the major part of which is for the circulating water pumps and the venting system [16].

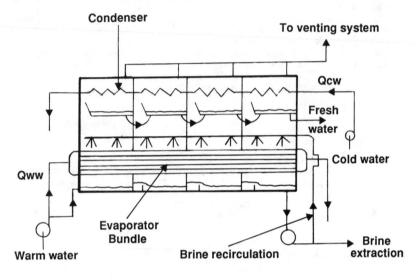

Figure 6.14 : Horizontal sprayed tubes
desalination process

### 6.4.3. Small- to Medium-Sized Desalination Units

The ocean thermal gradient could be used to run small- to medium-sized plants with the sole objective of fresh water production. Although one of the above desalination processes could be used, low cold- water consumption is an important selection criterion and the most appropriate solution appears to be the multieffect process [7]. The performances will depend on the site conditions and the process design (see Figure 6.15). Performance is increased by reducing the temperature difference between the effects and by reducing the head loss in the seawater exchangers.

Specific electrical consumption of the multieffect units with production capacity between 250 and 1000 m$^3$/day ranges from 6 kWh/m$^3$ down to 4 kWh/m$^3$. A good level of performance is achieved, for instance, by a unit with the following typical characteristics : available temperature difference of 22°C ; warm- to fresh-water ratio of 84 ; cold- to fresh-water ratio of 16 ; specific electrical energy consumption of 3.5 kWh/m$^3$.

Conventional technology of metallic tube-bundle heat exchangers can be used throughout. Because the units operate at low temperature and because of the reduced pressure differences between each effect, evaporators/condensers of the falling-film type have been designed, with exchange surfaces made of very thin (50 micrometers) and long (4 m) flexible nylon hoses about 30 mm in diameter [11]. Evaporation and condensation tests in the OTEC temperature range have been conducted on this kind of heat-exchange surface in a seawater test loop (CEA - French organism in charge of nuclear energy development - and IFREMER). These tests, run at thermal power levels from 10 to 20 kW, have shown that overall heat-exchange coefficients could reach typical values of 2.5 kW/m$^2$°C for a 2°C temperature difference between effects.

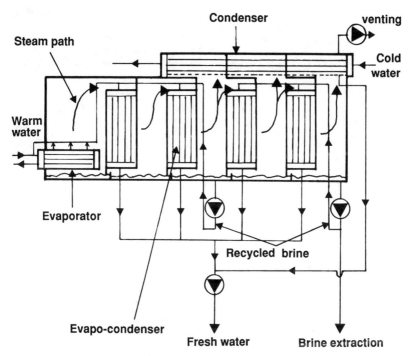

Figure 6.15 : Multi-effect desalination process

## 6.5.   Conclusions

Power was successfully generated by an OC-OTEC system in Cuba in 1929. From then until 1950, an audacious pioneering effort proved the viability of OC-OTEC and led to innovative designs, a number of which are still up to date, such as spout evaporators and upstream predeaerators. The next 25 years were a period of inactivity, for this and many other alternative energy systems. After the oil embargo, another period of intense activity, from the mid-1970s to about the mid-1980s, focused on OC-OTEC systems both in the United States and France.

Under the auspices of the U.S. Department of Energy, a very exhaustive assessment has been conducted on the various components of an OC-OTEC process. Extensive analytical models of OC-OTEC components were developed and validated with experimental data at SERI. Evaluation of the various possible designs was supported by a number of laboratory experiments at increasing scales, culminating in a series of tests with actual cold- and warm-seawater flows carried out at the Seacoast Test Facility in Hawaii.

In France, the OC-OTEC program has been managed by IFREMER and development activities have been pursued in cooperation with the industrial joint venture ERGOCEAN. The program objective was the design of a demonstration 5-$MW_e$ net power plant to be located in French Polynesia. In this project, the choice of component options was directed toward short-term available solutions. Some laboratory tests were conducted on the selected designs to ascertain their performances.

Since the mid-1980s, when energy costs began to decline, OTEC activities have been drastically slowed down. In France, the decision to build the 5-$MW_e$ pilot plant was postponed. Presently, the only active OC-OTEC program in the world is the U.S. program. It is hoped that the Net Power-Producing Experiment (NPPE) will soon be the first OC-OTEC demonstration plant of modern design to produce electricity since the short-lived Cuba experiment in 1929.

## 6.6.  References

1   Bharathan, D., Parsons, B. K., and Althof, J. A., "Direct-Contact Condensers for Open-Cycle OTEC Applications : Model Validation with Fresh-Water Experiments for Structured Packings," *SERI/TR-253-3108*, Golden, CO : Solar Energy Research Institute, May 1988.

2   Bharathan. D., and Penney, T., "Flash Evaporation from Turbulent Water Jets," *ASME Journal of Heat Transfer*, Vol. 106, 1984, pp.407-416.

3   Bharathan, D., Green, H.J., Link, H.F., Parsons, B.K., Parsons, J.M., and Zangrando, F., "Conceptual Design of an Open-Cycle Ocean Thermal Energy Conversion Net Power-Producing Experiment (OC-OTEC NPPE)", *SERI/TR-253-3616*, Golden, CO : Solar Energy Research Institute, July 1990.

4   Block, D. L., and Valenzuela, J. A., "Thermoeconomic Optimization of Open-Cycle OTEC Electricity and Water Production Plants," *SERI/STR-251-2603*, Golden, CO : Solar Energy Research Institute, May 1985.

5   Claude, G., "Power from the Tropical Seas," *Mechanical Engineering*, Vol. 52, No. 12, 1930, pp. 1039-1044.

6   Claude, G., "Sur l'Energie Thermique des Mers. La campagne de la **Tunisie** (On the Thermal Energy of the Sea. The campaign of the *Tunisie*)," *Comptes-rendus de l'Académie des Sciences*, Tome 200, No. 12, 1935.

7   Damy, G., and Marvaldi, J.,"Some Investigations on the Possibility of Using Ocean Thermal Gradient for Seawater Desalination," *Desalination*, Vol. 67, 1987, pp. 197-214.

8   Fournier, T., "Open-Cycle Ocean Thermal Energy Conversion: Experimental Study of Flash Evaporation," *Proceedings*, Oceans '85, San Diego, CA, Nov. 12-14, 1985, Vol.2, pp.1222-1229.

9   IFREMER, "OTEC, the French Program, a Pilot Electric Power Plant for the French Polynesia," *Internal Document*, Brest, France : Institut Français de Recherche pour l'Exploitation de la Mer, 1985.

10  Kreith, F., and Boehm, R.F., *Direct-Contact Heat Transfer*, Washington, D.C. : Hemisphere Publishing Corp., 1988.

11  Lauro, F., "Usine de Dessalement à Multiple-effet Utilisant des Surfaces d'Echanges en Matière Plastique (Multi-effect Desalination Plant Using Plastic Heat-exchange Surfaces)", *Proceedings,* 5th International Symposium on Fresh Water from the Sea, Alghero, Sardinia, May 1976, Vol. 2, pp. 261-268.

12  Lauro, F., "Système de Distillation de l'Eau de Mer à Partir des Rejets Thermiques (Seawater Distillation System from Waste Heat)", *Proceedings,* 6th Symposium on Fresh Water from the Sea, Las Palmas, Canary Islands, September 1978, Vol. 1, pp. 167-171.

13  Nisolle, M. L., "Utilisation de l'Energie Thermique des Mers (Les Problèmes de Fonctionnement) [Utilization of the Thermal Energy of the Sea (Operational Problems)]", Paris, France : Société des Ingénieurs Civils de France, pp.796-825.

14  Panchal, C.B., and Bell, K.J., "Theoretical Analysis of Condensation in the Presence of Noncondensable Gases as Applied to Open Cycle OTEC Condensers," *ASME Paper 84-WA/Sol-27*, 1984.

15  Parsons, B.K., and Link, H.F., "System Studies of Open-Cycle OTEC Components", *SERI/TP-253-2794*, Golden, CO: Solar Energy Research Institute. Presented at the Oceans'85 Conference, San Diego, CA., Nov. 12-14, 1985.

16  Rey, M., and Lauro, F., "Ocean Thermal Energy and Desalination", *Desalination*, Vol.39, 1981, pp. 159-168.

17  Saari, R., "Desalination by Very Low Temperature Nuclear Heat", *Nuclear Technology*, Vol. 38, 1978, pp. 209-214.

18  Salle, M. and Capestan, A., "Travaux Anciens et Récents sur l'Energie Thermique des Mers (Historical and Recent Investigations on the Thermal Energy of the Sea)," *La Houille Blanche*, No. 5, 1957, pp. 702-711.

19  University Study Teams, "Innovative Turbine Concepts for Open-Cycle OTEC," Final Report of Work Performed by Massachusetts Institute of Technology, Texas A&M University, University of Pennsylvania, *SERI/TR-253-3549*, Golden, CO: Solar Energy Research Institute, December 1989.

20  Vanier, C., and Masset J.F., "Accueil : Logiciel d'Aide à la Modélisation et à l'Optimisation de Systèmes Complexes Multi-paramètres (Welcome: Aiding Software for Modeling and Optimizing Multivariable Complex Systems)," *Association Technique Maritime et Aéronautique (ATMA)*, Paris, France, May 1987.

21  Westinghouse Electric Corporation, "100 MWe OTEC Alternate Power Systems, Vol. 1 : Technical Details," *Final Report*, Contract No. EG-77-C-05-1473 with U. S. Department of Energy, March 5, 1979.

22  Zangrando, F., Bharathan, D., Link, H., and Panchal, C. B., "Seawater Test Results of Open- Cycle Ocean Thermal Energy Conversion (OC-OTEC) Components," *Journal of Heat Transfer Engineering*, Vol. 11, No. 4, pp.44-53.

23  Zangrando, F., Bharathan, D., Green, H.J., Link, H.F., Panchal, C.B., Parsons, B.K., Parsons, J.M., Pesaran, A.A., "Results of Scoping Tests for Open-Cycle OTEC Components Operating With Seawater", *SERI/TR-253-3561*, Golden, CO : Solar Energy Research Institute, September 1990.

## List of Figures

## Nomenclature

| | |
|---|---|
| $C_p$ | specific heat |
| $h_{fg}$ | latent heat of vaporization |
| MW | molecular weight |
| pp | partial pressure |
| P | pressure, power |
| R | universal gas constant |
| Q | mass flow rate |
| $R_Q$ | ratio of warm- to cold-water mass flow rates |
| $R_T$ | ratio of turbine temperature difference to overall usable temperature difference |
| T | temperature |
| V | vent ratio |

## Subscripts

| | |
|---|---|
| c | condenser |
| cw | cold water |
| ev | evaporator |
| i | into component |
| o | out of component |
| r | resource |
| s | steam |
| sat | saturation |
| turb | turbine |
| u | usable |
| w | seawater, water |
| ww | warm water |
| 1,2 | condenser stages |

Greek

| | |
|---|---|
| $\delta, \Delta T$ | temperature difference |
| $\Delta h_{is}$ | isentropic enthalpy across the turbine |
| $\varepsilon$ | thermal effectiveness |
| $\eta_{TG}$ | efficiency of turbine/generator unit |
| $\Sigma \Delta P$ | sum of gas pressure drops |

# 7: Economics of Ocean Thermal Energy Conversion (OTEC)

Luis A. Vega, Ph.D. [1,2]

**ABSTRACT**

A straightforward analytical model is proposed to compare the cost of electricity produced either with OTEC or with petroleum or coal-fired plants. In the case of OTEC, when appropriate, the cost of electricity is estimated with credit for the desalinated water produced. The production cost of OTEC products are levelized over the life of the plant (nominal value: 30 years). Two generalized markets are considered: industrialized nations and smaller, less-developed island nations with modest needs. The model is used to establish scenarios under which OTEC could be competitive.

The scenarios are defined by two parameters: fuel cost, and the cost of fresh water production. In the absence of natural sources of fresh water, it is postulated that the cost of producing desalinated water from seawater via reverse osmosis (RO) be considered as the conventional technique. This approach yields a direct relationship between desalinated water production and fuel cost; and therefore, a scenario defined with one parameter.

It is determined that OTEC should only be considered as a system to produce electricity and desalinated water, because OTEC–based, mariculture operations and air–conditioning systems can only make use of a small amount of the seawater available; and therefore, could only impact small plants. The use of energy carriers (e.g.: Hydrogen, Ammonia) to transport OTEC energy generated in floating plants, drifting in tropical waters away from land, is determined to be technically feasible but requires increases in the cost of fossil fuels of at least an order of magnitude to be cost effective.

It is postulated that OTEC plants will be limited, by the relatively large diameter required for cold water pipes, to sizes of no more than 100 MWe-net (10 m diameter) in the case of floating plants and somewhat less (the value is a function of bathymetry or pipe length) for land-based plants. Furthermore, in the case of open cycle the plants will be limited by the low pressure turbine to 2.5 MWe-net modules or, for example, 10 MWe-net plants (arbitrarily, setting at four the number of modules per plant). Although the future rests in relatively large closed cycle OTEC floating plants, given the low level funding available for development of alternative energy, the first commercial plants will have to be 1 to 10 MWe land-based plants designed for the less-developed islands and funded by international aid agencies. The analysis shows that these, first generation, plants will have to produce electricity and desalinated water to offset the relatively higher cost of electricity; and, that their commercialization should be preceded by the installation of a demonstration plant of at least 1 MWe and 1,700 m³ to 3,500 m³, of desalinated water, per day production capacity. This demonstration plant would be used to obtain operational information and optimize the design of the first generation of commercial plants.

---

[1] Pacific International Center for High Technology Research (PICHTR), 711 Kapiolani Boulevard, Honolulu, Hawaii 96813.

[2] The views expressed in this chapter are those of the author and do not necessarily represent those of PICHTR.

It is determined that plants of at least 50 MWe capacity would be required for the industrialized nations; and, that if desalinated water is required to reach wider scenarios, it is proposed that a hybrid plant be used, based on the closed cycle for the electricity production and a second-stage, for desalinated water production, consisting of a flash (vacuum) evaporator and surface condenser. Closed cycle plants, without second-stage desalinated water production, are found to be cost effective if housed in floating vessels, moored or dynamically positioned a few kilometers from land, transmitting the electricity to shore via submarine power cables. The moored vessel could also house a hybrid OTEC plant and transport the desalinated water produced via flexible pipes. It is recommended that a floating 5 MWe and 7,500 m³/day demonstration plant be designed, installed and operated prior to the commercialization of plants of at least 50 MWe capacity.

## Background

The search for renewable sources of energy has resulted in the revival of a concept based on the utilization of the differences in temperature, $\Delta T$, between the warm ($T_W \approx 22°C$ to $29°C$) tropical surface waters, and the cold ($T_C \approx 4°C$ to $5°C$) deep ocean waters available at depths of about 1,000 m, as the source of the thermal energy required to vaporize and condense the working fluid of a turbine-generator system. This concept is referred to as Ocean Thermal Energy Conversion (OTEC).

There are two approaches to the extraction of thermal energy from the oceans, one referred to as "closed cycle" and the other as "open cycle." These approaches are described in Chapters 5 and 6 respectively, and briefly summarized here as an introduction to the analysis that follows. In the closed cycle, seawater is used to vaporize and condense a working fluid, such as ammonia, which drives a turbine-generator in a closed loop, producing electricity. In the open cycle, surface water is flash-evaporated in a vacuum chamber. The resulting low-pressure steam is used to drive a turbine-generator. Cold seawater is used to condense the steam after it has passed through the turbine. The open cycle can, therefore, be configured to produce fresh water as well as electricity.

The closed cycle was first proposed in 1881, by D'Arsonval in France, and was demonstrated in 1979, when a small plant mounted on a barge off Hawaii (Mini-OTEC) produced 50 kW of gross power, for several months, with a net output of 18 kW. This closed cycle plant was sponsored by private industry and the State of Hawaii. Subsequently, a 100 kW gross power, land-based plant was operated in the island nation of Nauru by a consortium of companies sponsored by the Japanese government. These plants were designed with public relations as the main objective and minimal operational data was obtained.

The open cycle concept was first proposed in the 1920's and demonstrated in 1930, off Cuba by its inventor, a Frenchman by the name of Georges Claude. His land–based demonstration plant was designed to resolve some of the ocean engineering issues common to all OTEC plants and, hopefully, to produce net electricity. This plant made use of a direct contact condenser; therefore, fresh water was not a by-product. The plant failed to achieve net power production because of a poor site selection (e.g., thermal resource) and a mismatch of the power and seawater systems; however, the plant did operate for several weeks.

An OTEC hybrid cycle, wherein electricity is produced in a first-stage (closed cycle) followed by water production in a second-stage, has been proposed as a means to maximize the use of the thermal resource available and produce water and electricity (Nihous, Syed and Vega, 1989). In the second-stage, the temperature difference available in the seawater effluents from an OTEC plant (e.g.: 12°C) is used to produce desalinated water through a system consisting of a flash evaporator and a surface condenser (basically, an open cycle without a turbine-generator). In the case of an open cycle plant, the addition of a second-stage results in doubling water production. Fresh water production with a flash-evaporator and surface condenser system was demonstrated in 1988 in a facility built by the U.S. Department of Energy at the Natural Energy Laboratory of Hawaii (NELH).

Floating vessels, approaching the dimensions of supertankers, housing factories operated with OTEC-generated electricity or transmitting the electricity to shore via submarine power cables have been conceptualized. Large diameter pipes suspended from these plantships extending to depths of 1,000 m are required to transport the deep ocean water to the heat exchangers onboard. The design and operation of these cold water pipes is a major issue that has been resolved (Vega and Nihous, 1988).

The proof-of-concept projects (i.e., Mini-OTEC, Nauru, Claude) demonstrated that both cycles are technically feasible and only limited, by the large diameters required for the cold water pipes, to sizes of no more than about 100 MWe (Vega, Nihous, Lewis, Resnick and Van Ryzin, 1989). In the case of the open cycle, due to the low-pressure steam, the turbine is presently limited to sizes of no more than 2.5 MWe.

Industry has not taken advantage of this information because, at present, the price of oil fuels and coal are such that conventional power plants produce cost-effective electricity. Moreover, the power industry can only invest in power plants whose design is based on similar plants with an operational record. Before OTEC can be commercialized, a prototypical plant must be built and operated to obtain the information required to design commercial systems and to gain the confidence of the financial community and industry. Conventional power plants pollute the environment more than an OTEC plant would and the fuel for OTEC is unlimited and free, as long as the sun heats the oceans; however, it is futile to use these arguments to convince the financial community to invest in an OTEC plant unless it has a proven operational record.

## Site Selection Criteria for OTEC Plants [3]

Except for closed basins, such as the Mediterranean and Red Seas, deep seawater flows from the polar regions: polar water, which represents up to 60% of all seawater, originates mainly from the Arctic for the Atlantic and North Pacific Oceans, and from the Antarctic (Weddell Sea) for all other major oceans. Therefore, $T_C$ at a given depth, approximately below 500 m, does not vary much throughout all regions of interest for OTEC. It is also a weak function of depth, with a typical gradient of 1°C per 150 m between 500 m and 1000 m. These considerations may lead to regard $T_C$ as nearly constant, with a value of 4°C at 1000 m.

---

[3] This section was written by Dr. G.C. Nihous.

Two facts require caution, however, during the OTEC site selection process: 1) OTEC is very sensitive to any loss of thermal resource, and 2) the Cold Water Pipe is a costly plant component. Consequently, variations in $T_C$ that appear to be small may have a drastic impact on the performance and/or the capital cost of the OTEC plant. For example, Pacific Ocean deep (1000 m) water at low latitudes is colder by about 1°C than Atlantic Ocean deep water; in the case of the East Coast of Africa, various phenomena including mixing with Red Sea outflow elevate the Indian Ocean water temperature (at 1000 m depth) to more than 6°C. As for the optimal depth at a given land-based OTEC site, seafloor bathymetry and topography play an important role and some degree of thermo-economic optimization is required; this point is discussed in more details in Nihous, Syed and Vega (1989), and Nihous, Udui and Vega (1989).

The other component of $\Delta T$ is $T_W$, the surface seawater temperature. In view of the above discussion regarding $T_C$, a desirable OTEC thermal resource of at least 20°C requires typical values of $T_W$ of the order of 25°C. Globally speaking, regions between latitudes 20°N and 20°S are adequate. Some definite exceptions exist due to strong cold currents: along the West Coast of South America, tropical coastal water temperatures remain below 20°C, and are often of the order of 15°C; a similar situation prevails to a lesser extent for the West Coast of Southern Africa. Moreover, $T_W$ varies throughout the year, and sometimes exhibits a significant seasonal drop due to the upwelling of deeper water induced by the action of the wind: such are the cases of the West Coast of Northern Africa in the Winter, where temperatures $T_W$ as low as 17°C are observed. More localized upwelling may occur, such as near the Horn of Africa in August, and in general, a careful OTEC site selection requires a comprehensive knowledge of local climate features inasmuch as they may affect $T_W$ seasonally.

The following summarizes the availability of the OTEC thermal resource throughout the World:

- Equatorial waters, defined as lying between 10°N and 10°S are adequate except for the West Coast of South America; significant seasonal temperature enhancement (e.g., with solar ponds) would be required on the West Coast of Southern Africa; moreover, deep water temperature is warmer by about 2°C along the East Coast of Africa.

- Tropical waters, defined as extending from the equatorial region boundary to, respectively, 20°N and 20°S, are adequate, except for the West Coasts of South America and of Southern Africa; moreover, seasonal upwelling phenomena would require significant temperature enhancement for the West Coast of Northern Africa, the Horn of Africa, and off the Arabian Peninsula.

The accessibility of deep cold seawater represents the most important physical criterion for OTEC site selection, once the existence of an adequate thermal resource has been established. Naturally, the distance from shoreline where water depths of the order of 1000 m may be found is far more critical for land-based plants than for floating plants, since it determines the length of the costly cold water pipe; in the case of a floating plant, the issue of cold seawater accessibility is only relevant inasmuch as a power cable, and, maybe, a small fresh water pipe, are needed to "export" the OTEC products to shore.

A valid way to assess cold seawater accessibility is to use a simple rule of thumb derived from the bathymetry of some of the potential OTEC sites in the World, namely that the 1000 m contour depth lie within 3000 m from shoreline. It should be emphasized that a study by Nihous, Udui and Vega (1989) demonstrated that the sensitivity of the OTEC production cost of electricity to detailed seafloor bathymetry is mostly pronounced for smaller plant sizes (1 to 10 MWe net power); for larger plants, a considerable economy of scale for OTEC seawater systems greatly reduces the importance of average seafloor slope.

The West Coast of Northern Africa or the East Coast of Africa, fall into the category of sites where warm water enhancement would be highly desirable because of a relatively adverse seafloor bathymetry (aside from the occurrence of a seasonal upwelling in the former case). The case of Australia is even worse: this country consists of a tectonic plate that extends far offshore, e.g., to the Great Barrier Reef, on its Northern tropical side. The Arabian Peninsula is also bordered by waters too shallow for OTEC to be practical.

Thus, physical factors affecting OTEC site selection, i.e., thermal resource and seafloor bathymetry, greatly restrict the number of desirable sites along the shoreline of major continents, unless some warm seawater temperature enhancement is possible. Most of the best, land-based, OTEC sites consist of island locations.

Finally, shoreline and hinterland topography, or extended shallow coral reefs require special attention. In the latter case, strong environmental concerns may not allow the building of any structure on the reef itself, although it represents a natural barrier against large breaking waves. Moreover, certain apparently favorable OTEC sites may be flanked by precipitous cliffs, or their hinterland may be very limited by rising mountain slopes.

The severe constraint of a favorable bathymetric profile, for the practical implementation of land-based OTEC technologies, would be relaxed to a considerable extent if the notion of floating OTEC plants were revived. In fact, the potential benefits of OTEC could only be recovered on a large scale through the development of an ambitious floating-plant program, following the initial experimental land-based OTEC phase.

Many other points must be considered when evaluating potential OTEC sites, from logistics to socioeconomic and political factors. One argument in favor of OTEC lies in its renewable character: it may be seen as a means to provide remote and isolated communities with some degree of energy independence, and to offer them a potential for safe economic development. Paradoxically, however, such operational advantages are often accompanied by serious logistical problems during the plant construction and installation phases: if an island is under development, it is likely to lack the infrastructure desirable for this type of project, including harbors, airports, good roads and communication systems.

Moreover, the population base should be compatible with the OTEC plant size: adequate manpower must be supplied to operate the plant, and the electricity and fresh water plant outputs should match local consumption in orders of magnitude. 1 to 10 MWe plants would generally suffice in most small Pacific islands (e.g., see IFREMER, 1985), whereas in the case of a populous and industrialized country like Taiwan, the largest feasible OTEC plants, up to 100 MWe, could be eventually considered (Shyu, 1989).

Since environmental protection has been recognized as a global issue, another important point to consider is the preservation of the environment in the area of the selected site, inasmuch as preservation of the environment anywhere is bound to have positive effects elsewhere. OTEC definitely offers one of the most benign power production technology, since the handling of hazardous substances is limited to the working fluid (e.g.: ammonia), and no noxious by-products are generated; OTEC merely requires the pumping and discharge of various seawater masses, which, according to preliminary studies, can be accomplished with virtually no adverse impact. This argument should be very attractive, for pristine island ecosystems, as well as for already polluted and overburdened environments. For example, the amount of $CO_2$ released from electricity-producing plants (expressed in gr of $CO_2$ per kWh) ranges from 1,000, for coal fired plants, to 700, for fuel-oil plants, while for closed cycle OTEC plants the value is less than 12 and for open cycle as much as 40 (Green and Guenther, 1989; Vega, 1981).

One major difficulty with OTEC is not of a technological order: OTEC is capital-intensive, and the very first plants, mainly because of their small size, will require a substantial capital investment. Given the prevailing low cost of crude oil, and of fossil fuels in general, the development of OTEC technologies is likely to be promoted by government agencies rather than by private industry. The motivation of governments in subsidizing OTEC may vary greatly, from foreign aid to domestic concerns.

For the former case, ideal recipient countries are likely to be independent developing nations. If these countries' economic standing is too low, however, the installation of an OTEC plant rather than direct aid in the form of money and goods may be perceived as inadequate help. In addition, political instability could jeopardize the good will of helping nations to invest.

For the latter case, potential sites belong to, or fall within the jurisdiction of, developed countries. A study performed by Dunbar (1981) identified ninety-eight nations and territories with access to the OTEC thermal resource (20° C temperature difference between surface water and deep ocean water) within their 200 nautical mile exclusive economic zone (EEZ). For the majority of these locations, the OTEC resource is applicable only to floating plants (arbitrarily assuming that the length of the cold water pipe for a land-based plant should not exceed 3,000 meters).

Dunbar's study, performed for the U.S. State Department, postulated a significant market potential for OTEC (i.e., 577,000 MWe of new baseload electric power facilities). Unfortunately, now as then, there is no OTEC plant with an operational record available. This still remains the impediment to OTEC commercialization.

The following list of potential sites is reproduced from Dunbar (1981):

| GEOGRAPHICAL AREA | MAINLAND | | ISLAND | |
| --- | --- | --- | --- | --- |
| AMERICAS | Mexico | Guyana | Cuba | Guadeloupe (FR) |
| | Brazil | Suriname | Haiti | Martinique (FR) |
| | Colombia | French Guiana (FR) | Dominican Rep. | Barbados |
| | Costa Rica | Nicaragua | Jamaica | Dominica |
| | Guatemala | El Salvador | Virgin Is. (US) | St. Lucia |
| | Honduras | Belize | Grenada | St. Kitts (UK) |
| | Panama | United States | St. Vincent | Barbuda (UK) |
| | Venezuela | | Grand Cayman (UK) | Montserrat (UK) |
| | | | Antigua (UK) | The Grenadines (UK) |
| | | | Puerto Rico (US) | Curacao (NETH) |
| | | | Trinidad & Tobago | Aruba (NETH) |
| | | | Bahamas | |
| AFRICA | Nigeria | Gabon | Sao Tome & Principe | |
| | Ghana | Benin | Ascension (UK) | |
| | Ivory Coast | Zaire | Comoros | |
| | Kenya | Angola | Aldabra (UK) | |
| | Tanzania | Cameroon | Madagascar | |
| | Congo | Mozambique | | |
| | Guinea | Eq. Guinea | | |
| | Sierra Leone | Togo | | |
| | Liberia | Somalia | | |
| INDIAN/PACIFIC OCEAN | India | Australia | Indonesia | American Samoa (US) |
| | Burma | Japan | Philippines | Trust Territories (US) |
| | China | Thailand | Sri Lanka | Northern Marianas |
| | Vietnam | Hong Kong (UK) | Papua New Guinea | Guam (US) |
| | Bangladesh | Brunei | Taiwan | Kiribati |
| | Malaysia | | Fiji | French Polynesia (FR) |
| | | | Nauru | New Caledonia (FR) |
| | | | Seychelles | Diego Garcia |
| | | | Maldives | Tuvalu |
| | | | New Hebrides (UK/FR) | Wake Is. (US) |
| | | | Samoa | Solomon Is. |
| | | | Tonga | Mauritius |
| | | | Cook Is. | Okinawa (JAPAN) |
| | | | | Wallis & Futuna Is. (FR) |
| | | | | Hawaii |

## OTEC Potential Market

There are at least two distinct markets for OTEC: (i) industrialized nations and islands; and, (ii) smaller or less industrialized islands with modest needs for power and fresh water.

The following global indicators are useful in relating the size of the plants considered herein with the needs of a community: (1) domestic water needs in developed nations are met with 100 gallons ($\approx$ 400 liters) per person per day [the United Nations uses a figure of 50 gallons ($\approx$ 200 liters) for countries under development]; in agricultural regions the use is 7 to 10 times larger; (2) the electrical power needs (domestic and industrial) of each 1,000 to 2,000 people are met with 1 MW in industrialized nations, while in less developed countries (LDCs) the needs of 5 to 15 times more people are met with 1 MW.

The small OC-OTEC plants considered in this chapter could be sized to produce from 1 to 10 MW electricity, and at least 450 thousand to 9.2 million gallons of fresh water per day (1,700 to 35,000 m3/day). That is, the needs of LDCs communities with populations ranging from 4,500 to as much as 100,000 could be met. This range encompasses the majority of less developed island nations throughout the world.

The larger CC-OTEC or hybrid cycle plants can be used in either market for producing electricity and water. For example, a 50 MW hybrid cycle plant producing as much as 16.4 million gallons of water per day (62,000 m3/day) could be tailored to support a LDC community of approximately 300,000 people or as many as 100,000 people in an industrialized nation. It is interesting to note that the state of Hawaii could be independent, of conventional fuels for the production of electricity, by utilizing the largest floating OTEC plants (50 to 100 MWe-net) for the larger communities in Oahu ($\approx$ 800,000 residents), Kauai, Maui and the Island of Hawaii ($\approx$ 100,000 residents), as well as smaller plants satisfying the needs of Molokai ($\approx$ 8,000 residents). Taiwan (ROC) could use several plants to meet additional requirements projected for the near future. The majority of the nations listed in the previous section could meet all their electricity and water requirements with OTEC.

To assess scenarios under which OTEC might be competitive with conventional technologies, in the production of electricity and water, a straightforward analytical model is developed. First, the capital cost for OTEC plants, expressed in 1990 $/kWe, is established assuming modest engineering development. The relative cost of producing electricity ($/kWh) with OTEC, offset by the desalinated water production, is then equated to the fuel cost of electricity produced with conventional techniques to determine the scenarios (i.e., fuel cost and cost of fresh water production) under which OTEC could be competitive. Inherent to this approach is the assumption that operation and maintenance costs are the same for OTEC and conventional plants of the same power capacity. No attempt is made at speculating about the future cost of fuel. It is simply stated that if a situation is represented by one of our scenarios, OTEC would be competitive.

For each scenario obtained, the cost of desalinated water produced from seawater by reverse osmosis (RO) is also given because this cost must be greater than the water production credit that OTEC requires to be cost effective. Once the cost effective scenarios are established, under this straightforward approach, a more rigorous economic analysis could be performed to model expected inflation and levelized costs (or alternatively, present worth). However, at this stage of development the approach followed here should suffice.

## OTEC Capital Cost Estimates and Production Rates

The range of capital costs for single stage OTEC systems is given in Figure 1 for land-based plants rated at nominal values of 1, 10, 50 MWe and 50 and 100 MWe plantships. The economy of scale is obvious. All published estimates for closed or open cycle land–based plants, given in 1990 dollars, fall within the range bounded by the two lines (e.g.: Electric Power Research Institute, 1986; IFREMER, 1985; Nihous et al, 1989; Shyu, 1989; Vega et al, 1989; SERI, Appendix D, 1990). The apparent change in slope at

10 MWe is caused by the use of only three plant sizes and should not be assigned any scale significance; however, it is a convenient way to indicate the upper limit recommended for open cycle plants. At the present level of development, the cost differential between cycles is within the accuracy of the estimates (± 20%). However, while closed cycle should only be limited by the practical size of the cold water pipe (≤ 10 m) to plants rated at ≤ 100 MWe for floating plants, and somewhat less for land-based plants; open cycle will be limited by the low pressure turbine to 10 MWe plants, consisting of four 2.5 MWe modules. In the case of plantships or floating plants only the costs projected, by the Year 2000, for closed cycle plants (Tables 8a and 8b) are shown.

The upper line also represents the present cost estimates with the lower line corresponding to the costs projected by the year 2000 after engineering development and the operation of demonstration plants that are scaled versions of the future commercial plants. These costs, in 1990 dollars are given in Tables 5, 6, 7 and 8a & 8b. The basis for the projected cost reductions are indicated in the tables. For example, work currently underway at ALCAN/MARCONI (e.g.: Johnson, 1989) complemented with the work previously performed by researchers from the Argonne National Laboratory (e.g.: SERI, Appendix D, 1990) indicates that the cost of surface condensers for open cycle, or the second-stage water production unit, and both the evaporator and condenser for the closed cycle should decrease from ≈ $215/m$^2$ to $100/m$^2$. All other cost reductions indicated in the tables should result from the operations of the demonstration plants. Future capital costs, corresponding to the lower line in Figure 1, are used throughout the discussion that follows.

Table 1 gives the estimates for 1 MW-net (nominal) open cycle plants with and without second-stage desalinated water production as well as a plant with a system including the use of 90 kg/s of 6°C cold seawater as the chiller fluid for a standard air-conditioning unit supporting a 300 ton load (≈ 300 rooms). For the purpose of this discussion, the 240 kWe of electricity displaced are considered as additional production, resulting in a total production of 10.1 x 10$^6$ kWh and an adjusted equivalent capital cost of 20,000 $/kW-net. The cost figures are expressed in 1990 dollars. These plants would be designed utilizing the state-of-the-art, bottom-mounted cold water pipe technology (i.e., 1.6 m diameter, high-density, polyethylene pipe). It is assumed that the 1 MW plants could be deployed some time after 1995. Their commercialization must be preceded by the installation of a demonstration plant of 1 MWe and 4,000 m$^3$, of desalinated water, per day production capacity.

Capital costs and production rates for land-based plants are summarized in Tables 2 and 3 for 10 MW open cycle plants, considered, at present, to be the maximum size for this cycle, and 50 MW closed cycle or, if water production is marketable, hybrid cycle plants. The design of the 10 MW open cycle plant would be scaled from the 1 MW demonstration plant with a new design for bottom-mounted cold water pipes (e.g., see Vega, Nihous, Lewis, Resnick and Van Ryzin, 1989). The commercialization of the 50 MW plants must be preceded by the design and operation of a 5 MWe closed cycle demonstration plant. These land-based plants would require the development of new bottom-mounted cold water pipes.

To consider the 50 MWe OTEC plantship moored or dynamically positioned 10 km offshore, in the discussions that follow, a capital cost of 4,600 $/kW-net is estimated

(Table 8a) for an electrical production of 380 x $10^6$ kWh (higher than for the land-based plants because of lower pumping power requirements). This cost is also given in 1990 dollars for a system to be deployed by the Years 2000 to 2005 assuming modest engineering development. These plants would be designed utilizing the methodology already available for cold water pipes suspended from a vessel (Vega and Nihous, 1988). The capital cost for the 100 MWe plant, corresponding to a 10 m diameter cold water pipe, would be 4,200 $/kW-net with an electrical production of 700 x $10^6$ kWh. The capital costs as a function of offshore distance are given in Table 8b.

## Conventional Production of Electricity

The thermal efficiency ($\eta$) of conventional steam power plants, fired with oil or coal, ranges from 32% to 34% and has been reported to be as high as 36%. The higher value will be assumed in this report. This implies that 36% of the heat added is converted to net work. Net work is defined as the difference between the output from the turbine-generator and the work required to run the plant.

The cycle efficiency for diesel fuel power plants used in Pacific Islands probably ranges from 25% up to 35%. A properly maintained and operated plant can reach efficiencies of up to 36%. This value will be assumed here to determine the work available from the heat added by the diesel fuel. This value is higher than realized in small steam power plants used for generation of electric power only. However, the cost of fuel suitable for the diesel generator is higher than the fuel suitable for the steam plant. Due to fuel availability, most island nations with small electricity requirements (a few MW) utilize diesel generators.

The convention followed in power plant technology, to express plant performance is to consider the heat added to produce a unit amount of net work. This parameter is called the heat rate (HR) of the plant and is usually given in Btu/kWh. Therefore, the heat rate is inversely proportional to the thermal efficiency, $\eta$ = 3413/HR (i.e., 1 kWh = 3413 Btu at 60°F), such that a thermal efficiency of 36% corresponds to a HR of 9500 Btu/kWh [herein common usage dictates the use of mixed units.]

The heating values of standard coal and fuel oil are 12,000 x (1 ± 0.17) Btu/lbm and 144,000 x (1 ± 0.04) Btu/U.S. gallon, respectively. Therefore, the fuel cost incurred in producing electricity, expressed in $/kWh, with an oil-fired plant is (within 6%): 1.6 x $10^{-3}$ times CB, the cost of a barrel (42 U.S. gallons) of fuel [9500 Btu/kWh/ (42 gallons/barrel x 144,000 Btu/gallon) = 0.0016 barrel/ kWh]. Therefore, at $18 per barrel, the fuel cost is 0.0288 $/kWh. The same expression will be used for diesel generators.

The 180 MW coal-fired plant under construction in Hawaii (Oahu) can be used to determine the capital cost for conventional steam power plants and the equivalent cost of coal. The plant will use Indonesian coal, with a baseline heat value of 12,500 Btu/lbm, to be delivered to for $2.25 per million Btu ($62 per metric ton) such that the fuel cost incurred in producing electricity with a thermal efficiency of 36% would be 0.021 $/kWh [9500 Btu/kWh x $2.25 / $10^6$ Btu]. This is equivalent to oil fuel cost of $13/barrel. The electric output will be sold to Hawaiian Electric Company under a 30-year contract.

The total capital cost of the project has been estimated to be $383.5 million; or $2,100/kW (AES Corporation, 1990).

## Conventional Production of Desalinated Water

The cost of producing fresh water from conventional desalination plants (i.e., Reverse Osmosis and Multistage Flash) ranges from about 1.3 to 2 $/m$^3$ for a plant capacity of 4,000 m$^3$/day to approximately 1 $/m$^3$ for a 40,000 m$^3$/day plant.

The energy (cost) used in multistage flash (MSF) distillation is in the form of heat, usually as low pressure steam, and the shaft power to drive pumps and other auxiliaries. Reverse osmosis (RO) plants require energy solely as shaft power from, for example, an electric motor. At present, RO is considered the technique of choice. It can be shown that, fresh water production by RO from seawater costs 0.049.CB, in $/m$^3$, where CB is the cost of a barrel (42 gallons) of fuel. This expression is used here to establish the desalinated water cost corresponding to a given fuel cost scenario.

## OTEC Production of Electricity

The following formula, proposed by the Electric Power Research Institute, is used to calculate the production cost of electricity p levelized over the assumed life for the OTEC plant (nominal value: 30 years):

$$p \ (\$/kWh) = (FC \times CC + OM \times G \times CR) \ / \ (NP \times CF \times 8760)$$

FC   :   annual fixed charge, taken as 0.10 (e.g.: government loan)
CC   :   plant overall investment capital cost, in $
OM   :   operation and maintenance yearly $ expenditures
G    :   present worth factor, in years, estimated value 20
CR   :   capital recovery factor, taken as 0.09
NP   :   net power production, in kW
CF   :   production capacity factor, chosen as 0.80
8760 :   number of hours in one year (CF.8760 =7,008)

The first term simply represents the payment for a fixed interest loan valued at CC, $, over a prescribed term expressed in hourly payments, where, the loan is for a plant rated at a power of NP, kW. The second term models the levelized cost of operating and maintaining the plant over the term.

For closed cycle plants, p is estimated with no credit taken for the sale of the fresh water by-product. For open or hybrid cycle plants, fresh water credit is obtained by multiplying the unit price by the yearly production and subtracting the result from the numerator of the expression given above. For the sake of completeness, costs estimated in this fashion are given in Tables 9 and 10 for unit prices of water at 0.4 $/m$^3$ and 0.8 $/m$^3$ respectively with the O&M expressed as a percentage of capital. It would be appropriate to levelize both the O&M and the unit cost of water (equivalent to multiplying these parameters by 1.8: the product of the capital recovery factor and the present worth factor as given above); however, in the case of prevailing unit price of water (0.4 $/m$^3$), the unlevelized costs given in Table 9 are within 10% of the levelized costs. In the case of the higher unit price of water (Table 10) levelizing both the O&M and the cost of water produces a dramatic difference. For example, in the case of the 1 MW with second stage, the cost is reduced from 0.25 to 0.19 $/kWh and for the 10MW with second stage results in a reduction from 0.08 $/kWh to less than zero. This is equivalent to a scenario of extremely high production costs for fresh water and to discuss it any further would only be speculative. Therefore, only unlevelized costs are considered in this Chapter.

These estimates illustrate the importance of the water revenue for the small plants (1 to 10 MW), especially with the unit price of water at twice the present prevailing rate (i.e., Table 10). The cost of electricity without water production is 0.3 $/kWh and 0.18 $/kWh for the single stage 1 and 10 MWe land-based plants as compared with 0.28 $/kWh and 0.14 $/kWh for the prevailing water cost (Table 9) and 0.25 $/kWh and 0.11 $/kWh for the higher water cost (Table 10). In the case of the larger plants water would be important only as a product that might be needed at a specific site (or under scenarios wherein the cost of conventional water production increases by factors of three to four times the prevailing value).

The capital cost estimates given above indicate that OTEC is a capital-intensive \technology. For example, the capital costs for oil-fired plants and coal-fired plants are less than $2,100/kW, as compared with the $10,700/kW and $6,000/kW given for the 10 MW and 50 MW plants in Tables 2 and 3. The 1 MW plant should be compared with diesel generators whose capital cost is less than 3% the cost of OTEC. However, OTEC incurs no fuel costs while conventional steam plants and diesel generators incur fuel costs. The levelized cost of OTEC electricity can be estimated from the equation given above; however, for the purpose of this report the capital cost of OTEC electricity, adjusted for the capital cost of the conventional technology (taken as $2,000/kW for oil or coal-fired plants and neglected for diesel generators) and the desalinated water by-product, will be compared with the fuel cost for conventional power plants to determine scenarios, given by the costs of electricity and water in a particular location, under which an OTEC plant of a given size could be cost competitive. Implicit in this approach is the assumption that O&M costs are similar for OTEC and conventional plants of the same power rating.

This approach can be formalized as follows for oil-fired plants:

$$(FC \times CC - WC \times PW) / (CF \times 8760 \times NP) < 1.6 \times 10^{-3} \times CB$$

Where,  FC, CC, CF and NP are defined above, and
WC  :  unit price of water, in $/m^3$
PW  :  yearly production of water, in m^3/year
CB  :  cost of a barrel of fuel (42 gallons), in $/barrel.

The production capacity factor (CF) is taken as 0.8 (80%) and the fixed charge for the capital (FC) as 0.1 (10%) and all values are expressed in present day costs. The water production of an open cycle plant is related to the amount of warm seawater utilized in the power cycle (i.e., between 0.4% and 0.5% of the warm seawater is flash evaporated in the process). As given in Tables 1, 2 and 3, this can be expressed as, PW = 130 x 10^6 gal/year/MW (or 50 x 10^4 m^3/year / MW-net). The second-stage described above would increase the water production in the case of an open cycle with second-stage by a factor of ≈ 2.2. In the case of the hybrid plant (i.e., flash evaporator/surface condenser downstream of closed cycle plant) the water production is equal to PW. Therefore, to determine the scenarios under which OTEC is competitive with oil-fired plants, the following expressions are used:

$$\frac{CC^*}{NP} = 110 \times CB + 5,000 \times WC, \qquad \text{for all cycles;}$$

and

$$\frac{CC^*}{NP} = 110 \times CB + 11,000 \times WC, \qquad \text{for open cycle with second-stage.}$$

\* For the 10 MW, 50 MW and 100 MW cases, \$2,000/kW are subtracted from the capital cost of OTEC to account for the capital cost of the conventional steam power plant. For diesel generators their capital cost can be neglected. For the closed cycle, the second term is omitted. The first term on the right hand side is accurate within 6% , and the second within 10%

The scenarios identified following this procedure are summarized in Table 4 for the capital cost and net power given in Tables 1, 2 , 3 and 8 for 1, 10 and 50 MW plants. Diesel fuel is considered at the 1 MW level, while less expensive oil fuel is used for the 10 MW and 50 MW cases.

## 1 MW Plants

This approach indicates that the 1 MW open cycle with second-stage water production (i.e., Table 1) would be competitive in a scenario given by a location where a high unit price of water, 1.6 \$/m³ (6 \$/kilogallon), and diesel fuel costs at \$45/barrel. The cost of producing desalinated water via RO would be 2.2 \$/m³ at this fuel cost. This scenario corresponds to conditions existing in certain less-developed Island nations with small populations. For example, in 1989 the cost of imported diesel fuel paid by the power companies was \$47/barrel in Western Samoa; \$50/barrel in the Kingdom of Tonga; and, \$25/barrel in Molokai.

The analysis indicates that small open cycle OTEC plants (Table 1) without second-stage water production could only be competitive under a scenario of diesel at \$93/barrel and the high unit price for the water, 1.6 \$/m³. This scenario does not appear likely. A 1 MW closed cycle plant (or open cycle without water production) would require a scenario where the diesel cost is \$165/barrel. The closed cycle plant with second-stage water production (hybrid plant) would require a scenario given by the high water cost, 1.6 \$/m³, and \$135/barrel.

It is interesting to note that the plant including a 300 ton AC system ($\approx$ 300-rooms hotel), described by Nihous, Syed and Vega (1989) and summarized in the last column of Table 1, would be competitive under a scenario given by \$45/barrel and 1.25 \$/m³ of water due to the additional revenue (electricity savings)generated by the use of 90 kg/s cold seawater as the chiller fluid for the air conditioning system. This amount of water amounts to only 3% of the water used for the 1 MWe OTEC plant. The 3,000 kg/s of cold seawater required for this size plant could support up to a 17,000 ton AC load. Therefore, the use of electricity savings from AC systems to offset the cost of OTEC electricity can only be consider for the small plants and is insignificant for the large plants discussed below. The use of cold seawater as the chiller fluid for AC systems represents a concept that is technically feasible and cost effective independently of OTEC.

## 10 MW Plants

For the 10 MW open cycle plant with second-stage water production, with a capital cost estimate of $14,700/kW (i.e., Table 2), a scenario given by $30/barrel of oil fuel and 0.85 $/m$^3$ of water is required. For the single stage OC-OTEC, with a capital cost of $10,700/kW a scenario given by $44/barrel of oil fuel and 0.8 $/m$^3$ of water is required. For the closed cycle system (or open cycle without water production) the scenario required is given by oil fuel at $ 80/barrel. For the hybrid plant (closed cycle with second stage) $79/barrel of oil fuel and 0.8 $/m$^3$ of water or $43/barrel of oil fuel and 1.6 $/m$^3$ of water are required.

These scenarios are plausible by the Year 2000 in a few small island nations. As indicated above, the capital cost for OTEC has been adjusted to account for the $2,000/kW capital investment for oil fuel plants. Once more, the additional water production makes the difference for OTEC for these relatively small plants. The cost of producing desalinated water via RO would be higher under all scenarios.

## 50 MW and 100 MW Plants

The land-based plant without water production summarized in Table 3 can be competitive under a scenario of fuel oil at $37/barrel. This is plausible by the Years 2000 to 2005. If water production is considered with a hybrid cycle plant, a scenario given by oil fuel at $49/barrel and unit price of water at $0.4/m$^3$; or another scenario wherein the unit price of water doubles and oil fuel is $31/barrel would be required. These scenarios might occur by the first decade of next century. The cost of RO water production is greater under both scenarios.

The plantship, housing a closed cycle 50 MW plant, could be competitive under a scenario of $23/barrel. For the 100 MW case the required cost is $20/barrel. If hybrid plants are considered, with the desalinated water transported to shore via flexible pipes, the scenarios required would be given by $23/barrel and $0.55/m$^3$ for the 50 MW plant and $20/barrel and $0.5/m$^3$ (or $24/barrel and $0.4/m$^3$) for the 100 MW plantship. The capital cost estimates used here are for plants deployed within 10 km from the shore line. The capital costs and the scenarios required for plants at offshore distances of 50 km , 100 km and 200 km are given in Table 8b. For example, for the 100 MW plantship positioned 50 km offshore a scenario given by $28/barrel would be required.

These scenarios are plausible in the majority of the sites listed above in the section entitled "Site Selection Criteria for OTEC Plants." The simple analysis presented in this chapter indicates that the future of OTEC rests in closed cycle floating plants that can also be configured to produce desalinated water as required.

## Co-Products of OTEC

The seawater needed for OTEC can also be used to support mariculture operations. The cold seawater contains large quantities of the nutrients required to sustain marine life. Organisms already grown in this environment include algae, seaweeds, shell fish and fin fish. Although a number of species have been identified as technically feasible, further work is required to identify cost effective culture methods for the available markets (Fast and Tanoue, 1988). The cold seawater can also be used as the chiller fluid for air-conditioning systems. A system based on this concept is presently utilized at NELH for one of the buildings.

In considering the economics of OTEC, it is appropriate to determine if multiple-product systems (e.g.: electricity, desalinated water, mariculture, AC systems) yield higher value by, for example, decreasing the equivalent cost of electricity. Unfortunately mariculture operations, as in the case of AC systems, can only use a relatively minute amount of the seawater required for OTEC systems. For example, the cold water available from a 1 MW OTEC plant could be used for daily exchanges of twenty-five 100m x 100m x 1m mariculture ponds, requiring at least 25 ha. Moreover, no mariculture operation requiring the use of the high-nutrient-deep-ocean water has been found to be cost effective. It is, therefore, recommended that OTEC be considered for its potential impact in the production of electricity and desalinated water and that mariculture and AC systems, based in the use of deep ocean water, be considered decoupled from OTEC.

A summary of the evaluation of co-products that the author has considered in conjunction with a 1 MW system is presented here for future reference and because the results were not published in the open literature. A mathematical model of the cold seawater utilization for mariculture and cooling applications was developed by Bhargava and Evans (1989). The California red abalone (*Haliotus rufescens*) and giant kelp (*Macrocystis spp.*) were chosen as the two species best suited for utilization of the cold seawater downstream of the OTEC plant. These species were selected for the model because they require seawater at approximately 15° C, abalone is a high value (sale price) product, and kelp is the naturally-grown feed for abalone. Dependence on outside sources for natural or artificial feed was, therefore, eliminated. Several other species considered (e.g., salmon) for OTEC-based, mariculture operations were not economically feasible because of the feed cost (Fast and Tanoue, 1988).

The mariculture operation was modeled based on a commercial farm operating at Keahole Point, Hawaii (i.e., Ocean Farms of Hawaii) buying cold seawater from the Natural Energy Laboratory of Hawaii (NELH). The operational cost in the form of a fee is incurred by the farm owner for the cold seawater resource. Likewise, associated capital costs are borne by the farm owner. Estimates on capital cost, electrical power consumption, annual farm production, and net annual revenue were made. Land available for the farm is nominally assumed to be 40,000 m² (4 hectares) in view of land availability restrictions in potential island markets.

The value-added benefit for OTEC is the annual fee that is paid by the farm for the cold seawater resource. In the example summarized below, the fee rate is nominally assumed to be that charged by NELH 666 $/ (yr-kg/s) [42 $/ (yr-gpm)]. With a cold seawater demand of 1700 kg/s (26,500 gpm) at 10.5 to 12° C., which is 53% of the amount discharged by the 1 MW OC-OTEC plant, the fee is estimated to be 1,132,200 $/year. The results obtained with this model can be summarized as follows:

| Species Selected | Kelp and Abalone |
|---|---|
| Farm Area | 4 Hectares |
| Cold Seawater Demand (OTEC Effluent) | 1,700 kg/s |
| Electrical Power Demand | 212 kW |
| Kelp Production | 1527 ton-wet/yr |
| Abalone Production | 65 ton-wet/yr |
| Capital Cost of Farm | 7.82 $M |
| Gross Revenue from Abalone Sales @ 40 $/kg-wet | 2.62 $M/yr |
| Cold Seawater Fee Paid to OTEC @ NELH Rate | 1.13 $M/yr |
| Net Profit (Loss) to Farm Owner | (0.41) $M/yr |

This analysis indicates that the fees charged by NELH would result in a loss of $410,000/year. A reduction of 50% in the fee charged by NELH for the cold seawater would yield a profit of $170,000/year (a return in capital investment of 2.2%). With the seawater available at no cost, the return in capital investment would be 9%.

Based on this study we concluded that the operation at Keahole Pt. could not be profitable. [Our conclusion was recently corroborated by the news media in Hawaii when it reported (January 1991) that the farm was bankrupt with monthly expenses at $200,000/month and revenues at $40,000/month].

The model for cooling was confined to a cold seawater-based, chiller system that would replace a refrigerant-based, chiller system. Chiller type air-conditioning (AC) systems are commonly used in hotels and large buildings. The chiller water temperature at inlet to the AC system has to be about 7.2° C for maximum human comfort which requires that the cold seawater temperature going to a counterflow, surface type, heat exchanger has to be 6.1° C or lower to meet the minimum pinch temperature requirement of 1.1° C. Therefore, the seawater for chiller application must be tapped upstream of the OTEC plant. Cold seawater is already being used for chiller water cooling by the Natural Energy Laboratory of Hawaii. The cooling capacity of the AC system is about 15 tons and the estimated reduction in electricity consumption is about 6000 kWh per month.

Estimates on capital cost, annual electrical power consumption, and annual costs were made. The electrical power requirement was found to be substantially lower for the cold seawater-based, chiller cooling system. For example, the electrical power saving for a 700-ton AC load (e.g., a 1,000-room hotel) is estimated to be 483 kW ($3.38 \times 10^6$ kWh).

The value-added benefit for OC-OTEC could be taken to be an annual fee paid by the AC system owner for the cold seawater resource. The cold water required by the nominally-sized chiller system is estimated to be 135 kg/s, which is $\approx$ 4% of the demand for the 1 MWe-net OC-OTEC plant discussed herein. The example considered can be summarized as follows:

| Cooling Application | Chiller Water Cooling |
|---|---|
| Electrical Power Saving | 483 kW |
| Cold Seawater Demand (OTEC Influent) | 135 kg/s |
| Cooling Load | 700 tons |

Capital Cost of Cooling System............................... 563.5 $K
Gross Savings for the Cooling System Owner
 @ Electricity Rate of 0.10$/kWh (Hawaii) ................. 338.0 $K/yr
Cold Seawater Fee Paid to OTEC @ NELH Rate ............  89.9 $K/yr
Net Savings for Cooling System Owner ..................... 135.3 $K/yr

The return in capital investment would be 24%, at an electricity charge of 0.10$/kWh.

In summary, the value-added benefits to OTEC from mariculture and chiller subsystems were assessed in terms of fees for the cold seawater resource. At the rate charged by NELH for the seawater resource, the mariculture operation was not profitable and the chiller system, using ≈ 4% of the cold seawater from a 1 MW plant, provided a 24% return in capital investment. It must, however, be noted that cold seawater fees from mariculture and chiller systems can only be significant for small OTEC systems.

In search of additional uses for OTEC seawater and the value-added benefits that might result, two additional studies were commissioned. Laws (1989) performed field experiments to determine the feasibility of growing *Gracilaria* in the OTEC effluent, for chemicals (e.g., agar) or methane production. The objective was to determine whether the required $CO_2$ bubbling could be eliminated by water exchange with the OTEC effluent. Unfortunately, Laws found that the availability of the OTEC effluent did not eliminate the need for an external source of $CO_2$ and that the production of agar or methane from *G. coronopilofia* is not cost effective.

In search for other aquatic plants (excluding seaweeds) of potential economic value that could benefit from the availability of the OTEC effluent, Weaver (1990) performed a study based on archival information and personal interviews to establish the chemical composition of aquatic plants; information on their mass-culture; and, to identify products with established or potential demand by industry (e.g., food, food–additive, chemical or pharmaceutical) including their economic value. Weaver considered: *Porphyridium, Diatoms, Dunaliella, Spirulina, Anabaena Azolla* and found that, although there exists potential for products extracted form these microalgae, there was no specific requirement for the OTEC effluent nor was there an identifiable benefit from it.

Based on these studies it can be stated that OTEC-based mariculture is in its formative years and not ready for commercialization, or transfer to nations under development. With the exception of the relatively small use of the cold seawater as AC chiller fluid, OTEC should be considered for its potential production of electricity and desalinated water.

## OTEC Energy Carriers

Several means of energy transport and delivery from OTEC plants deployed throughout the tropical oceans and distances exceeding those listed in Table 8b (> 200 km) have been considered (Konopka, et al, 1976). OTEC energy could be transported via electrical, chemical, thermal and electrochemical carriers. The submarine power cable has been considered above in the discussion of the 50 and 100 MW plantships (Tables

8a and 8b). The technical evaluation of non-electrical carriers leads to the consideration of hydrogen produced using electricity and fresh water generated with OTEC technology. The product would be transported in liquid form to land to be primarily used as a transportation fuel.

A 100 MWe-net plantship can be configured to include the following functions, in addition to supplying the fresh water required for the electrolysis and the crew:

- AC-DC Rectification    (97% Efficient)
- Electrolysis           (79% Efficient;
                         Specific power consumption 4.2 kWh/Nm$^3$ H$_2$
                         Distilled water consumption 0.886 liter/Nm$^3$ H$_2$
                         Auxiliaries Power 0.517 MW)
- Liquefaction           (97% Efficient, Power Consumption 11.3 kWh/kg–LH$_2$ ;
                         92.5% of GH$_2$ input is liquified)
- Onboard Storage        (98% Efficient)
- Ocean Transport        (91% Efficient; 1,600 km)
- Land Storage           (98% Efficient)

In this fashion, the 100 MWe of OTEC electricity yield 1298 kg/h of liquid hydrogen with an upper heating value of 39.41 kWh/kg corresponding to a thermal efficiency of 51%. The production cost of liquid hydrogen delivered to the harbor is given by the sum of the OTEC electricity component (0.069 $/kWeh for the 100 MWe plantship) plus the hydrogen production and delivery component. The estimated capital cost for the non-OTEC components is $290,000,000 (Konopka, et al, 1976 and Wurster, et al, 1990). For the economic parameters used in the estimate of the cost of OTEC electricity (e.g.: fixed rates, O&M) the cost, given in 1990 dollars, per million Btu is 40 $/MBtu for the electricity component and 31 $/MBtu for the hydrogen subsytem for a total cost for the product delivered to the harbor of 71 $/MBtu (3413 Btu/kWh). This is equivalent to gasoline (@0.125 MBtu/ US gallon) priced at $9/gallon or crude oil (@ 5.8 MBtu/barrel) priced at $412/barrel. Considering the lowest capital cost and the highest system thermal efficiency projected in the literature the cost for the product delivered to the harbor is reduced to 47 $/MBtu corresponding to gasoline at $6/gallon or crude oil at $273/barrel.

The situation is similar for the others energy carriers considered in the literature. For example, for liquid ammonia the cost of the delivered product (liquid hydrogen) is 88% of the cost given above; for methylcyclohexane the cost is 96%; and, for metal hydrides 108%. For a land-based OTEC system a reduction of 30% in the cost of the end product is achieved (i.e.: 50 $/MBtu instead of 71 $/MBtu for the baseline case) by increasing the production of liquid hydrogen by 12% and reducing the capital cost on the non-OTEC components by 50%. All carriers considered yield costs higher that those estimated for the submarine power cable (Table 8b) Therefore, the only energy carrier that is cost effective for OTEC energy is the submarine power cable.

## Externalities in the Production and Consumption of Energy

At present, the external costs of energy production and consumption are not considered in the determination of the charges to the user. Considering all stages of generation, from initial fuel extraction to plant decommissioning, it has been determined that no energy technology is completely environmentally benign. The net social costs of the different methods of energy production is a topic under study. Hubbard (1991) has published the range of all estimates reported in the literature for the costs due to: corrosion, health impacts, crop losses, radioactive waste, military expenditures, employment loss, subsidies (tax credits and research funding for present technologies). The sum of all estimates yields a range of 78 to 259 billion dollars per year. Excluding costs associated with nuclear power, the range is equivalent to adding from $85/barrel to $327/barrel. As minimum, consider that the costs incurred by the military, in the USA, to "safeguard" oil supplies from overseas is at least $15 billion or equivalently $23.5/barrel. Accounting for externalities might eventually help the development and expand the applicability of OTEC, but in the interim the scenarios that have been identified herein should be considered.

It is interesting to note that discussions, in the public sector, regarding taxing the electricity produced by non-renewable resources at a rate of $0.02/kWh is equivalent to adding $13 per barrel (42 gallons) of fuel to the cost of electricity production. This tax would not suffice to make small OTEC plants cost effective in industrialized nations; however, in the case of the larger plants this tax would work in favor of OTEC.

## Development Requirements

The analysis presented herein indicates that there is a market for OTEC plants that produce electricity and water. Industrialized nations and islands, as well as island nations under development, could make use of 50 MW to 100 MW closed cycle or hybrid plants housed in plantships (e.g.: Tables 8a and 8b). A few less developed or smaller islands could use of open cycle plants with second-stage additional water production (e.g., Tables 1 and 2).

However, operational data must be made available, for example, to establish production factors and plant reliability. This data can only be obtained by building and operating demonstration plants scaled from the commercial-size plants listed in Tables 1, 2, 3, and 8. A plan aimed at achieving the development of OTEC under the scenarios discussed in this chapter is summarized in Table 11 (SERI, Appendix D, 1990).

The potential locations considered to develop this plan with the aim of 2100 MWe installed by the Year 2010 were (Note: that 95% of the plants would be based on the closed cycle):

| | | |
|---|---|---|
| Small Pacific Islands ..... | 100 MW | (Open Cycle) |
| Taiwan...................... | 400 MW | (Closed or Hybrid Cycle) |
| Oahu ....................... | 200 MW | (Closed or Hybrid Cycle) |
| Hawaii ..................... | 50 MW | (Closed or Hybrid Cycle) |
| Molokai..................... | 10 MW | (Open Cycle) |
| Kauai ...................... | 40 MW | (Closed or Hybrid Cycle) |
| Philippines................. | 400 MW | (Closed or Hybrid Cycle) |
| Indonesia .................. | 200 MW | (Closed or Hybrid Cycle) |
| India ...................... | 200 MW | (Closed or Hybrid Cycle) |
| Puerto Rico ................ | 200 MW | (Closed or Hybrid Cycle) |
| Gulf of Mexico............. | 300 MW | (Closed or Hybrid Cycle) |

It is interesting to note that the earlier study by Dunbar (1981) considered a potential market of 577,000 MW of new baseload, electric power facilities.

## Conclusions

The identification of the scenarios—given by the cost of fuel oil and the production cost of water—under which OTEC systems are cost competitive indicates that closed cycle OTEC plants of at least 50 MWe and as much as 100 MWe capacity must be considered. If desalinated water is required, a hybrid system configured with second stage water production is applicable. The lowest costs correspond to plantships deployed close to the shoreline. An exception is found under scenarios with high costs for the fuel oil and the conventional production of water, where small (1 to 10 MWe) land-based OC-OTEC plants, with second stage for additional water production, are cost effective. These scenarios correspond to small markets found in only a few island nations. These conclusions must be confirmed by designing, constructing and operating demonstration plants scaled from the commercial size plants. A 5 MWe plantship with second stage water production is an appropriate size for the demonstration plant, scaled from the 50 to 100 MWe plants. Likewise a 1 MWe OC-OTEC demonstration plant with second stage additional water production must be considered as a scaled version of the plants for the small island market.

The small OC-OTEC plants considered in this chapter could be sized to produce from 1 to 10 MW electricity, and at least 450 thousand to 9.2 million gallons of fresh water per day (1,700 to 35,000 m³/day). That is, the needs of LDCs communities with populations ranging from 4,500 to as much as 100,000 could be met. This range encompasses the majority of less developed island nations throughout the world.

The larger CC-OTEC or hybrid cycle plants can be used in either market for producing electricity and water. For example, a 50 MW hybrid cycle plant producing as much as 16.4 million gallons of water per day (62,000 m³/day) could be tailored to support a LDC community of approximately 300,000 people or as many as 100,000 people in an industrialized nation. It is interesting to note that the state of Hawaii could be independent, of conventional fuels for the production of electricity, by utilizing the largest floating OTEC plants (50 to 100 MWe-net) for the larger communities in Oahu ($\approx$ 800,000 residents), Kauai, Maui, and the Island of Hawaii ($\approx$ 100,000 residents), as well as smaller plants satisfying the needs of Molokai ($\approx$ 8,000 residents). Taiwan (ROC) could use several plants to meet additional requirements projected for the near future. The majority of the nations listed in the previous section could meet all their electricity and water requirements with OTEC.

An assessment of the state of the art and evaluation of potential developments reveals that the only energy carrier that is cost effective for OTEC energy is the submarine power cable and that, with the exception of the relatively small use of the cold seawater as AC chiller fluid, in conjunction with the 1 MWe land-based plants, OTEC should only be considered for its potential production of electricity and desalinated water.

Accounting for externalities in the production and consumption of energy might eventually help the development and expand the applicability of OTEC, but in the interim the scenarios that have been identified herein should be considered. Conventional power plants pollute the environment more than an OTEC plant should and the fuel for OTEC is unlimited and free, as long as the sun heats the oceans; however, it is futile to use these arguments to convince the financial community to invest in OTEC plants without an operational record.

## References

AES Corporation, "AES Closes $383.5 Million Project Financing for Construction of Power Plant in Hawaii," Press Release, April 3, 1990. Contact: P. Hanrahan, Arlington, Virginia (703) 358-0506.

Bhargava, A., Evans, D.E., "OC-OTEC Computer Model Incorporating Electricity and Desalinated Water Production, and Cold Seawater Utilization for Mariculture and Cooling Applications," PICHTR Publication, Honolulu, Hawaii, August 1989.

Dunbar, L.E., "Market Potential for OTEC in Developing Nations," Proceedings, 8th Ocean Energy Conference, Washington, D.C., June 1981.

Electric Power Research Institute, "Ocean Energy Technologies: The State of the Art," EPRI AP-4921, November 1986.

Fast, A.W., Tanoue, K.Y., "OTEC Aquaculture in Hawaii," University of Hawaii SEAGRANT College Program, Contribution No. 59 (UNIHI–SEAGRANT–MR–89–01), November 1988 (177 pages).

Green, H.J., Guenther, P.R., "Carbon Dioxide Release from OTEC Cycles," Proceedings, International Conference on Ocean Energy Recovery, Honolulu, Hawaii, November 1989. Published by the American Society of Civil Engineers.

Hubbard, H.M., "The Real Cost of Energy," *Scientific American,* April 1991.

IFREMER, Centre de Brest, Projet E.T.M., "Situation des Etudes d'Avant-Projet de la Centrale E.T.M. de Tahiti au 31.03.85," Report #DIT/SP/ETM 85.312 MG/ES, August 1985.

Konopka, A.J., Talib, A., Yudow, B., Biederman, N., "Optimization Study of OTEC Delivery Systems Based on Chemical–Energy Carriers," Institute of Gas Technology (IGT), December 1976, ERDA Report No. NSF–C1008 (AER–75–00033).

Laws, E.A., "The Feasibility of Growing Gracilaria in OTEC Effluent," University of Hawaii at Manoa, prepared for PICHTR, January 1989.

Nihous, G.C., Syed, M.A., and Vega, L.A., "Conceptual Design of a Small Open–Cycle OTEC Plant for the Production of Electricity and Fresh Water in a Pacific Island," Proceedings, International Conference on Ocean Energy Recovery, Honolulu, Hawaii, November 1989. Published by the American Society of Civil Engineers.

Nihous, G.C., Udui E. and Vega, L.A., "Preliminary Evaluation of Potential OTEC Sites: Bathymetry and Feasibility," prepared by PICHTR for the Commonwealth of the Northern Mariana Islands, Contract #C549994-01 with the U.S. Department of Energy, Territorial Assistance Program, October 1989.

Solar Energy Research Institute (SERI), et al, "The Potential of Renewable Energy: An Interlaboratory White Paper," SERI/TP–260–3674, March 1990. Prepared for the Office of Policy, Planning and Analysis, U.S. Department of Energy in support of the National Energy Strategy. Available from: National Technical Information Service. [Appendix D. *Ocean Energy Technologies*]

Shyu, Chuen-Tien, "Topography of the Eastern Taiwan Coastal Areas," Proceedings of the International Workshop on Artificial Upwelling and Mixing in Coastal Waters, pp. 124–132, Keelung, Taiwan, June 1989.

Vega, L.A., Bailey, S., "Assessment of OTEC-Based Mariculture Operations," Proceedings, 8th Ocean Energy Conference, Washington, D.C., June 1981.

Vega, L.A., Nihous, G.C., "At-Sea Test of the Structural Response of a Large Diameter Pipe Attached to a Surface Vessel," Proceedings, Offshore Technology Conference, No. 5708, pp. 473-480, Houston, Texas, May 1988.

Vega, L.A., Nihous, G.C., Lewis, L. Resnick, A., and Van Ryzin, J., "OTEC Seawater Systems Technology Status," Proceedings, International Conference on Ocean Energy Recovery, Honolulu, Hawaii, November 1989. Published by the American Society of Civil Engineers.

Weaver, E.C., "Use of OTEC Water for Cultivating Aquatic Plants of High Economic Value," prepared for PICHTR, March 1990.

Wurster, R., Malo, A., "The Euro–Quebec Hydro–Hydrogen Pilot Project," Proceedings of the 8th World Hydrogen Energy Conference, Honolulu, Hawaii (held in July 1990), Vol. 1, p. 59.

Capital Costs Applicable to Open Cycle (≤ 10 MWe) and Closed Cycle (≤ 100 MWe) Plants

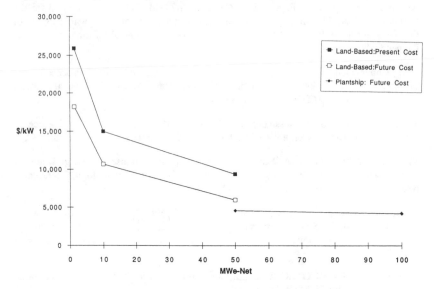

Figure 1.   **Capital Cost for Single Stage OTEC Plants**

| PLANT NOMINAL SIZE: | 1 MW; OC-OTEC | 1 MW; OC-OTEC with 2nd-Stage | 1 MW; OC-OTEC with 2nd-Stage and 300 ton A-C |
|---|---|---|---|
| **PRODUCTION:** | | | |
| Electricity | $9.5 \times 10^6$ kWh | $8.8 \times 10^6$ kWh | $8.4 \times 10^6$ kWh plus the A-C equivalent |
| Water | 0.45 MGD (1,700 m$^3$/day) | 1.06 MGD (4,000 m$^3$/day) | 1.06 MGD (4,000 m$^3$/day) |
| A-C Electricity Equivalent | -- | -- | 300 ton A-C @ 0.8 kW/ton Equivalent to $1.7 \times 10^6$ kWh |
| **CAPITAL COST:** | 18,200 $/kW-net | 23,000 $/kW | 20,000 $/kW (adjusted) |
| **YEAR DEPLOYED:** | 1995 | 1995 | 1995 |

NOTE: Estimates are from Table 5 for the Capital Cost expected after engineering development in compressors and turbine efficiency as well as for cost reductions in surface condensers from $215/m$^2$ to $100/m$^2$. State-of-the-art seawater piping systems are used.. The estimate for the plant with a 300 ton A-C is for electricity production savings due to the use of 90 kg/sec of cold seawater (3% of the total cold water required for power cycle) as a chiller fluid for a standard A-C unit.

Table 1.   **Capital Cost Estimates ($/kW-net) for 1 MW Land-Based Plants in 1990 Dollars**

| PLANT NOMINAL SIZE: | 10 MW; OC-OTEC | 10 MW; OC-OTEC with 2nd-Stage |
|---|---|---|
| **PRODUCTION:** | | |
| Electricity | $70 \times 10^6$ kWh | $63 \times 10^6$ kWh |
| Water | 4 MGD (15,000 m$^3$/day) | 9.2 MGD (35,000 m$^3$/day) |
| **CAPITAL COST:** | 10,700 \$/kW-net | 14,700 \$/kW |
| **YEAR DEPLOYED:** | 2000 | 2000 |

NOTE: Estimates are from Table 6 for the Capital Cost expected after cost reductions in surface condensers from \$215/m$^2$ to \$100/m$^2$; engineering development resulting in improved turbines and vacuum compressors and new cold-water-pipe technology.

Table 2. **Capital Cost Estimates (\$/kW-net) for 10 MW Land-Based Plants in 1990 Dollars**

| PLANT NOMINAL SIZE: | 50 MW; CC-OTEC | 50 MW; Hybrid Ammonia Power Cycle with 2nd-Stage Water Production |
|---|---|---|
| **PRODUCTION:** | | |
| Electricity | $336 \times 10^6$ kWh | $280 \times 10^6$ kWh |
| Water | n/a | 16.4 MGD (62,000 m$^3$/day) |
| **CAPITAL COST:** | 6,000 \$/kW | 9,400 \$/kW |
| **YEAR DEPLOYED:** | 2000 to 2005 | 2000 to 2005 |

NOTE: Estimates are from Table 7 adjusted as described in Table 2.

Table 3. **Capital Cost Estimates (\$/kW-net) for 50 MW Land-Based Plants in 1990 Dollars**

| NOMINAL NET POWER (MWe) | TYPE | SCENARIO REQUIREMENTS | SCENARIO AVAILABILITY |
|---|---|---|---|
| **1**<br><br>*(Table 1)* | Land-Based OC-OTEC with 2nd-Stage additional Water Production | • $45/barrel of diesel<br>• $1.6/m³ water | South Pacific Island Nations by Year 1995 |
| **10**<br><br>*(Table 2)* | Land-Based (as above) | • $30/barrel of fuel oil<br>• $0.85/m³ water | American Island Territories and other Pacific Islands by Year 2000 |
| **50**<br><br><br><br><br>*(Table 3)* | Land-Based Hybrid (ammonia power cycle with Flash Evaporator downstream)<br><br>Closed-Cycle Land-Based | • $49/barrel of fuel oil<br>• $0.4/m³  water<br>—or—<br>• $31/barrel<br>• $0.8/m³ water<br>• $37/barrel | Hawaii, if fuel or water cost doubles by Year 2000 |
| **50 & 100**<br><br><br>*(Table 8)* | Closed-Cycle Plantship ( 50 MW )<br><br>Closed-Cycle Plantship (100 MW ) | • $23/ barrel<br><br>• $20/ barrel | Present |

- OC-OTEC  limited by turbine technology to 2.5 MW modules or 10 MW plant (with four modules)
- CC-OTEC  or Hybrid (water production downstream of closed-cycle with flash evaporator)

Table 4.   **OTEC Market Penetration Scenarios**

| | 1 MWe OC-OTEC (1990) | 1 MWe OC-OTEC (1995) | COMMENTS |
|---|---|---|---|
| • STRUCTURE | 4,400 | 3,700 | Structural Improvements after demonstration plant |
| • SEAWATER SYSTEM (Pipes and Pumps) | 12,300 | 7,600 | Improvement in SWS technology |
| • POWER SYSTEMS<br>- Heat Exchangers | 3,500 | 2,000 | Surface Condenser reduced from $215/m$^2$ to $100/m$^2$ |
| - Turbine-Generator and Vacuum Compressor | 3,700 | 3,200 | Increases in efficiencies for T-G and Vacuum Compressor systems augment net power from 1156 to 1356 kW |
| • OTHER | 2,000 | 1,700 | Engineering Development |
| TOTAL | 25,900 | 18,200 | |

NOTE: Second Stage Water Production Unit doubles desalinated water production, decreases net power by 100 kW and capital costs increase by $4.3 M (e.g., 23,000 $/kW by Year 1995).

Table 5.  **Potential Reduction in Capital Cost ($/kW) by Year 1995, Expressed in 1990 Dollars, for a Land-Based 1 MWe OC-OTEC Plant**

| | 10 MWe OC-OTEC (1990) | 10 MWe OC-OTEC (2000) | COMMENTS |
|---|---|---|---|
| • STRUCTURE | 1,500 (10%) | 1,500 (14%) | --- |
| • SEAWATER SYSTEM (Pipes and Pumps) | 6,000 (40%) | 3,700 (34%) | New CWP Technology ( d > 1.6 m ); |
| • POWER SYSTEMS<br>- Heat Exchangers | 3,500 (23%) | 2,000 (19%) | Surface Condenser reduced from $215/m$^2$ to $100/m$^2$ |
| - Turbine-Generator and Auxiliaries | 2,500 (17%) | 2,000 (19%) | Engineering Development results in improved turbine and vacuum compressor |
| • OTHER | 1,500 (10%) | 1,500 (14%) | --- |
| TOTAL | 15,000 | 10,700 | |

NOTE: Second Stage Water Production Unit doubles desalinated water production, decreases net power by 1,000 kWe and augments capital investment by $25 M (e.g., 14,700 $/kW by Year 2000)

Table 6.  **Potential Reduction in Capital Cost ($/kW) by Year 2000, Expressed in 1990 Dollars, for a Land-Based 10 MWe OC-OTEC Plant**

| | 50 MWe CC-OTEC (1990) | 50 MWe CC-OTEC (2000) | COMMENTS |
|---|---|---|---|
| • STRUCTURE | 1,500  *(18%)* | 1,500  *(25%)* | --- |
| • SEAWATER SYSTEM (Pipes and Pumps) | 2,400  *(29%)* | 1,700  *(28%)* | New CWP Technology |
| • POWER SYSTEMS | | | |
| - Heat Exchangers | 2,500  *(30%)* | 1,200  *(20%)* | Condenser reduced from $215/m$^2$ to $100/m$^2$ |
| - Turbine-Generator | 1,200  *(15%)* | 1,000  *(17%)* | Engineering Development |
| • OTHER | 600  *(8%)* | 600  *(10%)* | --- |
| TOTAL | **8,200** | **6,000** | |

NOTE: For the Hybrid Plant (Second Stage Water Production), net power is decreased by 8,000 kW and capital cost is incremented by $60 M (e.g., 9,400 $/kW by Year 2000)

**Table 7.  Potential Reduction in Capital Cost ($/kW) by Year 2000, Expressed in 1990 Dollars, for a Land-Based 50 MWe CC-OTEC Plant**

| | MOORED 50 MWe CC-OTEC (1990) | MOORED 50 MWe CC-OTEC (2000) | COMMENTS |
|---|---|---|---|
| • VESSEL / MOORING AND POWER CABLE | 1,800  *(26%)* | 1,200  *(26%)* | Engineering Development |
| • SEAWATER SYSTEM (Pipes and Pumps) | 800  *(12%)* | 600  *(13%)* | Engineering Development |
| • HEAT EXCHANGERS | 2,500  *(36%)* | 1,200  *(26%)* | Condenser and Evaporator reduced from $215/m$^2$ to $100/m$^2$ |
| • TURBINE-GENERATOR | 1,200  *(17%)* | 1,000  *(22%)* | Engineering Development |
| • OTHER | 600  *(9%)* | 600  *(13%)* | --- |
| TOTAL | **6,900** | **4,600** | |

**Table 8a.  Potential Reduction in Capital Cost ($/kW) by Year 2000, Expressed in 1990 Dollars, for Floating 50 MWe CC-OTEC Plants**

| OFFSHORE DISTANCE | MOORED 50 MWe CC-OTEC (2000) | MOORED 100 MWe CC-OTEC (2000) | SCENARIO REQUIRED FOR 100 MWe PLANT |
|---|---|---|---|
| 10 km | 4,600 | 4,200 | $ 20 / barrel |
| 50 km | 5,500 | 5,000 | $ 28 / barrel |
| 100 km | 6,600 | 6,000 | $ 37 / barrel |
| 200 km | 8,900 | 8,100 | $ 55 / barrel |

Table 8b.  **Capital Cost ($/kW) for Single Stage Floating Plants, as a Function of Offshore Distance Projected by Year 2000 After Engineering Development and the Operation of Demonstration Plant**

| NOMINAL PLANT CF= 80% | Production Electricity kWh | Production Water m^3/day | Capital Cost $/kW-net | Fixed Charge | O&M % of Capital | Electricity Cost with Water Production Credit $/kWh |
|---|---|---|---|---|---|---|
| 1 MW OC OTEC Land Based | 9.50E+06 | 1,700 | 18,200 | 10% | 1.7% | 0.28 |
| 1 MW OC OTEC Land Based 2nd Stage | 8.80E+06 | 4,000 | 23,000 | 10% | 1.7% | 0.32 |
| 10 MW OC OTEC Land Based | 7.00E+07 | 15,000 | 10,700 | 10% | 1.5% | 0.14 |
| 10 MW OC OTEC Land Based 2nd Stage | 6.30E+07 | 35,000 | 14,700 | 10% | 1.5% | 0.16 |
| 50 MW CC OTEC Land Based | 3.36E+08 | 0 | 6,000 | 10% | 1.5% | 0.10 |
| 50 MW CC OTEC Land-Hybrid | 2.80E+08 | 62,000 | 9,400 | 10% | 1.5% | 0.12 |
| 50 MW CC OTEC Floater | 3.80E+08 | 0 | 4,600 | 10% | 1.5% | 0.08 |
| 50 MW CC OTEC Floater-Hybrid | 3.20E+08 | 62,000 | 7,200 | 10% | 1.5% | 0.09 |

Table 9.  **Cost of OTEC Electricity with Desalinated Water Credit at $0.4/m^3 ($1.5/1000 gallons)**

| NOMINAL PLANT CF= 80% | Production Electricity kWh | Production Water m^3/day | Capital Cost $/kW-net | Fixed Charge | O&M % of Capital | Electricity Cost with Water Production Credit $/kWh |
|---|---|---|---|---|---|---|
| 1 MW OC OTEC Land Based | 9.50E+06 | 1,700 | 18,200 | 10% | 1.7% | 0.25 |
| 1 MW OC OTEC Land Based 2nd Stage | 8.80E+06 | 4,000 | 23,000 | 10% | 1.7% | 0.25 |
| 10 MW OC OTEC Land Based | 7.00E+07 | 15,000 | 10,700 | 10% | 1.5% | 0.11 |
| 10 MW OC OTEC Land Based 2nd Stage | 6.30E+07 | 35,000 | 14,700 | 10% | 1.5% | 0.08 |
| 50 MW CC OTEC Land Based | 3.36E+08 | 0 | 6,000 | 10% | 1.5% | 0.10 |
| 50 MW CC OTEC Land-Hybrid | 2.80E+08 | 62,000 | 9,400 | 10% | 1.5% | 0.09 |
| 50 MW CC OTEC Floater | 3.80E+08 | 0 | 4,600 | 10% | 1.5% | 0.08 |
| 50 MW CC OTEC Floater-Hybrid | 3.20E+08 | 62,000 | 7,200 | 10% | 1.5% | 0.06 |

Table 10.   **Cost of OTEC Electricity with Desalinated Water Credit at $0.8/m^3 ($3/1000 gallons)**

| YEARS | PROJECT | FUNDS REQUIRED |
|---|---|---|
| **1990–1995** | A) • Reduce the cost of Surface Condensers;<br>• Develop Low Pressure Steam Turbines rated at about 2.5 MWe; and,<br>• Develop Large Diameter Cold Water pipes for Land–Based Plants (> 1.6 m). | $ 10M |
| | B) OC-OTEC / 2nd-stage Demonstration Plant (1 MWe / 3,500 m³/day) using state-of-the-art | $ 30 M |
| **1995–2000** | C) Deploy Land-Based Plants Optimized from (B) (Total 5 MWe / 17,500 m³/day) | International Banking / Aid Community |
| | D) Hybrid Land-Based Demonstration Plant (5 MWe / 7,500 m³/day) using newly developed CWP | $ 75 M (optional) |
| | E) CC-OTEC Plantship Demonstration Plant (5 MWe) using existing technology | $ 60 M |
| **2000–2005** | F) Deploy several Land-Based Plants in Pacific and Asia. Optimized from (D) | Private |
| | G) Deploy several 50 to 100 MWe Plantship in Tropical Waters Optimized from (E) | Private |
| **2005–2010** | H) Provide Projected Power and Water Increase in small Pacific Islands, and part thereof in Oahu, Taiwan, Philippines, etc., and Plantships (e.g.: Cumulative Deployed Power : 2100 MWe) | Cumulative Capital Investment $13 Billion (for 2100 MWe) |

Table 11.   **OTEC Development Program Required**

8: State of the Art in Wave Power Recovery
A.Douglas Carmichael[1]
Johannes Falnes[2]

## Abstract

Considerable developments have taken place in the last decade or so in the field of wave energy. Much of this research and development effort has been sponsored by government agencies, mainly in Europe and Scandinavia. Several efficient configurations of energy converters have been designed and tested at model scale, and some have been operated in the sea. Theoretical models of many types of wave energy converters have been published, and these analyses have facilitated a greater understanding of the processes that occur in practical systems, and have been an invaluable aid to the designer and inventor. The pace of developments has slowed in recent years but new wave energy converters are continuing to be studied.

## Historical Survey

An intensive research and development study of wave energy conversion began after the dramatic increase in oil prices in 1973. Reviews of much of this research are available in monographs (McCormick 1981, Shaw 1982, Claeson et al. 1987), conference proceedings (Count 1980a, Berge 1982, Evans and Falcão 1985, Duckers 1990) and recent review papers (Hagerman and Heller 1988, Salter 1989). Other reviews (Stahl 1892 and Leishman and Scobie 1976) have provided historical surveys of wave energy devices.

An operating wave energy system was constructed by Bouchaux-Praceique in France in the early twentieth century. In this system the rise and fall of the water surface in a special chamber communicating with the ocean, provided compressed air to drive a turbo-generator (McCormick 1981, p. 4). Masuda, a Japanese inventor, who has been working on wave energy devices for more than forty years, developed, in the 1960-s, a small air turbine system that could be placed on a buoy to provide electricity for navigation lights. More than one thousand of these units have been sold worldwide, and some

---

[1]Professor, Dept. of Ocean Engineering, Massachusetts Institute of Technology, Cambridge, MA 02139.

[2]Professor, Division of Physics, Norwegian Institute of Technology, University of Trondheim, Norway.

of them have been in operation for 20 years (Claeson et al. 1987, p.218).

Salter (1974) described a very efficient wave energy conversion device, later termed the Salter Duck, that was capable of absorbing more than 80% of the energy from waves over a range of wave frequencies. This very encouraging performance was measured on small models in a wave tank. Other inventors have since developed and tested many configurations of devices, including floating structures, pneumatic chambers, and devices having oscillating and moving parts, both on the surface and beneath the ocean. Some of these converters had good performance with efficiencies and general performance characteristics similar to the Salter Duck (Count 1980b).

Following Salter's experiments, Budal and Falnes (1975, 1977), Evans (1976), Newman (1976), Mei (1976) and Mei and Newman (1979) made theoretical studies of two-dimensional and three-dimensional wave-energy converters (WECs). The theoretical analyses were subsequently extended to the case of arrays of three-dimensional WECs (Budal 1977, Evans 1979, Falnes 1980).

A rather intense research and development program was carried out between 1974 and 1982 in the United Kingdom, with government funding. Government and private support of research programs also began before 1980 in Norway, Sweden and the U.S.. Funding of the research programs declined during the 1980-s in most of these countries, while in other countries, India and Portugal, some activities in wave power research and development were intensified (Raju et al 1991, Falcão 1990). Some small scale experiments and theoretical analyses have also been conducted in other countries, including Belgium, Canada, China, Denmark, France, Ireland and USSR (Lewis 1985, Li and Guo 1985, Liang et al 1991, Zossimov 1990). Coastal demonstration units were constructed during the 1980-s in Japan, Norway, and UK.

In the work funded by the British Government, experimental investigations and engineering evaluations were made of many different devices. No sea tests of full scale devices were conducted, and the support of wave energy research and development was reduced in 1982, and largely curtailed in July 1985, after spending more than 15 million pounds, about 24 million dollars (Davies et al. 1985). With industrial support, the British Government continued to fund modest developments of a device called SEA Clam for island communities. Since 1987 the British Government has given funds to support a group from Queen's University, Belfast, in the construction of a demonstration prototype unit on the west coast of Scotland. There is a current reevaluation of the previous work on the predicted cost of producing electricity from wave energy.

The Norwegian Government has supported research and development work on wave energy. Several different concepts have been evaluated, and two prototype devices were installed, in 1985, on the west coast of Norway. In two other Scandinavian countries, Denmark and Sweden, wave power devices have been tested at sea.

In the U.S. support from government funds has been at a modest level. However, industrial and private organizations have promoted several wave energy devices and concepts through preliminary design and model testing.

In Japan, Masuda continued working on wave energy. The Japan Marine Science and Technology Center supported Masuda's design of a large floating platform named "KAIMEI" with pneumatic power conversion. In early 1980 power from KAIMEI was delivered for a short period to the Japanese grid. During the 1980-s several prototype units of different types were tested at sea.

In Portugal, theoretical analysis and experimental work have been conducted since 1978 to develop the technology necessary for wave energy conversion. The work is continuing, and it is planned to install a pneumatic prototype device similar to one of the units placed on the west coast of Norway.

The five countries Canada, Ireland, Japan, UK, and the U.S. joined in the work of the International Energy Agency in the field of wave energy. This research and development effort has been conducted aboard the Japanese wave-absorbing platform "KAIMEI".

## Classification of Wave Energy Converters (WECs)

In the patent literature there are more than a thousand different proposals for the utilization of wave energy (McCormick 1981, p.4), and there are many ways of describing and classifying them. Several wave energy devices are illustrated on Figure 1. They may be described in terms of their location, theoretical attributes, general arrangement, or energy use.

Location

A wave energy converter may be placed in the ocean in various possible situations and locations. It may be floating or submerged in the sea offshore, or it may be located on the shore or on the sea bed in relatively shallow water. All of these four categories were represented by at least one WEC device analyzed in the British wave energy program around 1980.

A converter on the sea bed may be completely submerged, it may be surface-piercing or it may, as in the case

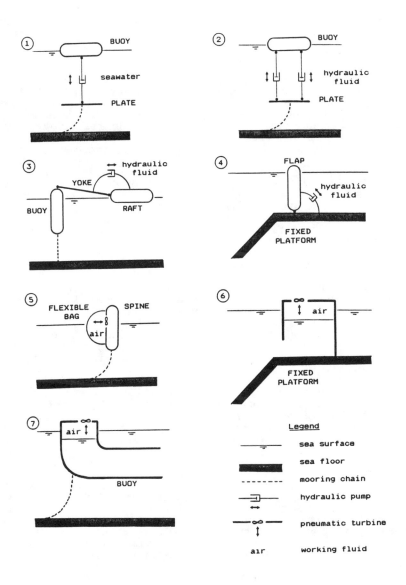

**Figure 1. Diagram Illustrating Several Types of Wave Energy Devices. (SEASUN Power Systems, 1988)**

of the Japanese prototype KAIYO, be a converter system placed on an offshore platform. However, apart from wave-powered navigation buoys, most of the prototypes constructed so far have been placed at or near the shore. An advantage with such systems is that operation and maintenance are relatively easy. Drawbacks are that the available wave power may be smaller than offshore, and the civil-engineering work is difficult on a wave-exposed shore. For offshore floating WECs the situation is different, operation and maintenance are more difficult, but construction (in a shipyard) may be relatively easy. Submerged WECs are even less accessible for maintenance, but they are less exposed to corrosion and to extreme wave forces.

It has not yet been determined which location is the best for a WEC. One may envisage hybrids of the various systems, for example, a device, not far offshore, may pump sea water into an elevated basin on land where a hydroelectric converter utilizes the water as it flows back to sea. There may be concern with the potential for ecological damage from such configurations.

Point Absorbers, Attenuators, and Terminators

WECs have also been classified (mainly by theoretical hydrodynamicists) according to their size and orientation. Devices which are very small compared to a typical wave-length, have been termed point absorbers. Since the power rating of a point absorber is typically a few hundred kilowatts, a large power plant would consist of hundreds or perhaps thousands of such units, which are dispersed in a very long and relatively narrow array along the coast.

The counterpart of point absorbers are elongated floating structures of a length which is comparable to or larger than one wave-length. If they are aligned roughly in the direction of wave propagation, they are commonly termed attenuators. If they are placed parallel to the prevailing wave front, they are termed terminators. Theory indicates that the maximum absorbed power occurs when the device is about one wavelength long, Evans (1982). The predicted absorbed power is equivalent to the incident power in a wave crest width corresponding to approximately 0.6 of a wavelength for an attenuator, and 1.2 wavelengths for a terminator; this depends on the configuration and operation of the devices. Theoretical analyses therefore indicate that the performance of the ideal attenuator would be somewhat less than the performance of the ideal terminator of equivalent length.

Types of Converted Energy and Energy Use

Other ways to classify various WECs may be based upon different kinds of mechanical energy conversion or different forms of the use of the converted useful energy.

The primary conversion of wave energy to some other kind of mechanical energy may be of different types. In oscillating water column (OWC) devices the energy is pneumatic, where air turbines use the flow of air, resulting from the pumping action of the oscillating water in an air chamber. Alternatively, oscillating water may be rectified through water turbines by means of valves. With another system, waves filling an elevated basin, supply potential energy to run a hydroelectric plant with sea water.

Many proposed devices utilize relative motion between bodies, where hydraulic cylinders and hydraulic motors are used to transmit and convert energy. Typical nineteenth-century inventions had mechanical transmission, through racks, pinions, cog wheels etc.

Converted wave energy may be used for various purposes. In addition to electricity production for delivery to a grid or for local use, converted wave energy may be used for desalination of sea water, for water pumping, for heating or cooling, or for the propulsion of vessels. In some cases wave-power conversion may be combined with the attenuation or the reduction of waves, to increase the weather window for beach activities or for offshore construction operations.

## General Principles for Conversion of Wave Power

Theoretical analyses of wave energy conversion have demonstrated how a unit absorbs energy from ocean waves. The important feature is that in response to incident waves, the device generates waves to cancel the passing and/or reflected waves. The analyses also indicate that for maximum power conversion, the velocity should be in phase with the exciting force; this condition occurs when the device is at resonance.

Generally it can be said that a good wave absorber must be a good wave-maker. Evans (1980) in a paper describing theoretical analyses of wave energy devices states "In this form it is clear that a good wave absorber is a body which has the ability when making waves, to concentrate the wave energy in a narrow sector rather than distribute the energy over all angles". Waves can be generated by an oscillating body, alternatively, waves may be produced by an oscillating water column (OWC) in a fixed chamber having an opening to the sea. It is also possible to enclose the water by blocking the chamber opening with an elastic or flexible membrane (as in several British inventions), which can oscillate under wave action. In such a case it is not necessary to have water in the chamber; it can simply be replaced by air, which oscillates in pressure, in step with the flexible membrane.

The ability to concentrate the direction of wave generation rather than distribute the energy over all angles can be obtained with an asymmetric device and is discussed in a later section. The Salter Duck has this configuration, it can generate waves effectively in the upstream direction but not downstream.

Resonance and Phase Control

It is an advantage to design a WEC with at least one resonance period in the range of periods where the wave energy is most abundant, typically between 5 and 15 seconds. If the so-called resonance bandwidth can be made large, then the phase may be approximately at the optimum value for wave periods significantly off resonance. However, a broad bandwidth requires a large structure, which means that relatively large masses of steel or concrete have to be used in constructing the WEC. A different approach is to save steel and concrete by designing a smaller structure, and compensating for the rather narrow resonance bandwidth by applying some form of control system to obtain optimum phase ("phase control", Budal and Falnes 1982).

Wave Power

The wave power per unit wave front in deep water is given by:

$$J = \rho(gH)^2 T/32\pi = \alpha H^2 T \tag{1}$$

for regular (sinusoidal) waves. In this expression $\rho$ is the mass density of water, $g$ the acceleration of gravity, $T$ the period of the wave, and $H$ the wave height (the vertical distance from wave trough to wave crest), and the coefficient $\alpha = 0.98$ kW/sm$^2$. The wavelength $L$ is related to the period through the equation

$$L = gT^2/2\pi = 1.56T^2, \text{ m} \tag{2}$$

For the representative values, $T = 10$ s and $H = 2$ m; these expressions give $J = 39$ kW/m and $L = 156$ m.

For irregular waves, where the height and the period vary from one wave to the next wave, the wave power transport per unit frontage is

$$J = \alpha_J H^2_{mo} T_J \tag{3}$$

where $\alpha_J = \alpha/2 = 0.49$ kW/sm$^2$ in deep water, and where the significant wave height $H_{mo}$ and the energy transport period $T_J$ are defined by means of the so-called wave spectrum. In most cases $H_{mo}$ is very close to the average wave height of the largest 1/3 of observed waves in a record, and $T_J$ is approximately 10 percent longer than the zero-upcrossing period $T_z$ defined as the average time interval between successive crossings of the mean water level as the water moves in the upward direction.

Maximum Absorbed Power

     Since the 1970-s it has been known (from the
above-mentioned theoretical studies) that there are upper
limits for the fraction of the energy of a wave that can be
absorbed by WECs. A single line of point absorbers or a
two-dimensional terminator cannot absorb more than half of the
incident wave energy if it is a symmetric system, that is, if
it is able to radiate large waves of equal magnitude in the
"upstream" direction (i.e. towards the incident waves) and the
"downstream" direction. However, it may be possible to absorb
all the energy of an incident wave, if the system is sufficiently non-symmetric in its ability to radiate waves in
opposite directions. Salter (1974) came rather close to this
ideal condition with his laboratory experiments on the Duck.

     The maximum power which may be absorbed by a heaving
axisymmetric point absorber is

$$P_{max} = \rho/2(gT/4\pi)^3 H^2 = (L/2\pi)J \qquad (4)$$

which is the incident power of a wave front of width equal to
the wavelength divided by $2\pi$. This width may be termed the
"absorption width". Early experimenters, not being aware of
this relationship, were surprised by measuring absorption
widths larger than the physical width of the model device. An
alternative to the term "absorption width" is "capture width".
This latter term is sometimes also used for the width corresponding to the useful power, which is the absorbed power
minus the power loss from friction and other dissipative
effects.

     In practice, the theoretical upper limit for the
absorbed power may be approached only if the wave is so small
that the maximum power handling capacity of the WEC does not
limit power take-up. Another condition for maximum power
absorption is that the WEC produces a water displacement
oscillation of optimum phase and optimum amplitude, in order
to obtain an optimum destructive interference between waves
radiated from the WEC and waves otherwise present in the sea.
Optimum phase is achieved when the oscillating system is in
resonance with the incident wave. The amplitude is optimum
when the loading of the oscillating system is adjusted to a
level where the absorbed power and the radiated power are the
same magnitude.

Maximum Power/Volume

     In the previous section the theoretical maximum
ratio of the power absorbed by a WEC and the incident natural
wave power was discussed. Another theoretical upper limit,
derived for a heaving point absorber (Budal and Falnes 1980),
relates the maximum power output to the physical size of the
device. The analysis indicates that the ratio of the absorbed

power to the oscillating volume displacement amplitude, should be less than

$$(P/V)_{max} = (\pi/4)\rho gH/T \qquad (5)$$

Conditions for obtaining this optimum are, first, that the oscillation phase is at the optimum value (phase control or resonance), and secondly, that the loading of the oscillating system is adjusted to a level where the radiated power is negligible compared to the absorbed power. This latter condition is in contrast with the corresponding optimum condition for obtaining the maximum absorbed power from the available energy in the ocean, as considered earlier. In the present case the optimum loading is much larger, but only a relatively small proportion of the available wave energy in the ocean is absorbed. The implication of this expression is that there may be better economy in minimizing the size of the point absorber for a selected power level, and exploiting only a portion of the available wave energy in the ocean, rather than converting as much of it as is technically possible.

Hence, it is important to consider the limit, given by equation (5), when designing WECs. Notice that the maximum value of P/V can be obtained only when the volume, V, approaches zero, a condition which is not practical with a WEC. It means however, that the larger the volume, V, the smaller the upper limit for P/V. With a representative ocean wave of height H = 2 m and period T = 10 s, equation (5) gives $(P/V)_{max}$ = 1.6 kW/m$^3$, whereas most of the WECs considered have corresponding figures in the range of 0.1 - 1 kW/m$^3$. Terminators and attenuators have figures in the lower part of this range and phase controlled point absorbers in the upper part. Most of the existing prototype WECs have been designed without consideration of the limits given by equation (5).

Importance of Wave Properties in Designing WECs

In a real sea the waves are neither regular in wave period nor uniform in wave height and statistical observations of real seas have been made at various ocean sites. The results from such studies are used to provide predictions of the wave energy, according to equation (3). The characteristics of waves have been measured around the oceans of the world for more than 100 years by oceanographers and mariners. Wave measurements have also been made over periods of several years at particular sites using electrical measuring devices. From these data it is possible to provide estimates of available natural wave energy in various ocean regions around the world.

Variability of Ocean Waves

The annual average wave energy may vary from one year to the next, and there are also seasonal variations. The

tion of wave energy. In the northern oceans of the world, the average wave energy for a winter month is usually several times the mean value for a summer month. In the U.S. this is out of phase with electrical loads because of summer air conditioning, but in Europe and other regions, the electrical loads are typically higher in the winter.

In addition to the seasonal variations of wave energy, there are changes that occur with shorter time scales. The average wave energy can vary by a factor of ten from one week to the next, and the wave energy during one storm can be five times higher than the mean value for the week that the storm occurs. On the time scale of minutes, or fractions of minutes, the wave energy in groups of large amplitude waves may be fifty times the wave energy between such wave groups. For these reasons it may be desirable to have some form of energy storage with wave energy conversion, with a storage capacity of at least a few minutes, to keep the electrical generator spinning.

Reflection, Refraction and Focussing

Ocean waves may be reflected, refracted or diffracted by objects in the ocean. Complete or partial reflection occurs when waves strike a solid object, such as a wall or a WEC device. Waves are refracted by changes in wave speed introduced by variations in the depth of the water, or by submerged objects. The refraction of ocean waves (approximately) obeys Snell's Law (from optics), and the phenomenon may be used to focus or defocus waves. Diffraction occurs when waves are partially shielded by a wall or by a floating object, provided it is not small. Waves entering a narrow opening into a harbor are diffracted after passing through the entrance.

In addition to natural focussing, mentioned above, artificial focussing of waves may be utilized in the collection of wave power. This can be achieved, for example, by using shaped, submerged bodies, whereby waves are concentrated to a relatively small, focal region. The performance of an ideal focussing device depends on its size, and, in principle, all the energy entering the device can be focussed on a small region where the wave energy converter would be placed. The performance of such devices varies with the wave period, but can be made to accommodate the expected predominant period of the waves. The performance also depends on the direction of the waves. Underwater structures to produce artificial focussing by refraction are expected to be large.

Open Ocean and Coastal Wave Climates

When waves become very steep they are unable to sustain their structure and spill forward, and are said to "break". While steep waves break and recover in deep water, in shallow water they tend to continue breaking and eventually

in shallow water they tend to continue breaking and eventually dissipate their energy as turbulence and thence to heat. Large breaking waves are destructive (Weigel 1964), and in shallow water, they can carry rocks and debris into wave energy conversion devices, and may damage moorings and cables.

Wave power is usually larger in open oceans than in seas, and in regions which are in some way shielded from the waves of the open ocean. For instance, the mean power transport is smaller in the North Sea than in the Atlantic Ocean off the west coast of the British Isles.

The mean wave power transport is generally reduced as the coast is approached. With decrease in water depth, energy is lost by bottom friction, percolation, non rigidity of the bottom, and more significantly, by wave breaking (Weigel 1964). Moreover, wind blowing from the land does not create big waves near the coast. When there is wave action in shallow water, the bottom topography can cause the refraction of waves, perhaps producing defocussing in some regions, and natural wave amplification (focussing) in other areas ("hot spots"). This depends on wave frequency as well as wave direction. An average, taken over all frequencies and all directions, may indicate regions of the sea, which are better locations for wave-power devices than other areas.

The mean power output, and hence, the economical income, is related to the average wave power at the site, while the design criteria, and hence the necessary investment in construction, are related to the extreme wave situation. Hence, it is advantageous to choose a location where extreme waves are reduced relative to the average waves (Hagerman 1985).

## Wave Energy Conversion Technology

Many wave energy devices have been invented but only a small proportion have been tested and evaluated, furthermore, only a very few have been tested at sea, in ocean waves. The technology described in this section of the report is therefore largely based on published model tests and engineering evaluations. Much of the technology comes from British, Japanese, and Norwegian sources, with contributions from the U.S. and other countries.

British Wave Energy Conversion Devices

The development of wave energy conversion was supported intensely by the British Government from 1974 to 1982. Many different devices were examined, and detailed studies were conducted on about a dozen concepts. Much of the performance information was measured on small models of 1/60-1/100 scale size, in wave tanks, and some tests were also conducted in lakes and sheltered coastal sites on larger scale models, of about 1/10 full size. A team of engineering consultants was given the task of assessing the power predictions, engineering viability, and the cost of producing electricity on a common basis for a number of the promising devices.

The British wave energy converters were optimized for a site off the Scottish island of South Uist, where the average incident wave power was measured to be about 48 kW/m and the mean wave period was approximately 8 seconds. Some of the technical and economic data remain proprietary, but much has been published (Davies 1985). The devices developed and evaluated in the United Kingdom include a wide range of configurations of wave energy converters designed for relatively large outputs. For the economic evaluation, clusters of several hundreds of these devices were assumed to provide outputs of about 2,000 MW (2GW) for 5% of the year. Descriptions of two of these units are presented here. The third British device is a more recent design.

Salter Duck. This was the earliest of the high efficiency wave energy devices. The Duck is a cam shaped device that oscillates about a spine connecting many devices. The spine is long enough to span several waves and provides (more-or-less) a steady frame of reference for power extraction. The design uses the precession of gyroscopes mounted in the nose of the oscillating Duck to drive a high pressure hydraulic system. This design has the Ducks mounted in pairs on sections of articulated spine. The joints in the spine have hydraulic rams which are controlled to regulate the flexing of the spine, and to provide optimum power extraction. The rated power of a Duck 38 m (125 ft) long, with a width of 22 m (72 ft) on a 14 m (46 ft) diameter spine was 2.25 MW, at the South Uist site. More recently a smaller device was proposed, the Solo Duck, which, if moored by tension legs, can manage without a spine, and without gyros, Figure 2, at the cost of mechanical penetrations of the casing. The necessary torque reaction could be obtained through a moored torque arm (Salter 1989).

SEA Clam. The device developed by Sea Energy Associates has some similarities with the Duck. The converter has a floating concrete spine with a number of flexible bags attached to one side, and moored at approximately 55 degrees to the incident wave direction, Figure 3. The action of the waves causes air to flow from the bags into the hollow spine through a self-rectifying air turbine. When the trough of the wave arrives the air flows back from the spine to the flexible bags through the same turbine. The predicted average output of the Sea Clam is about 12% of the incident wave power corresponding to the length of the device. The rated power of a convertor 290 m (951 ft) long, and 13 m (43 ft) beam is 10 MW, and the average power at the South Uist site is predicted to be approximately 1.72 MW. The inventor, Bellamy (1985), has indicated that the basic spine structure of the Clam, wrapped

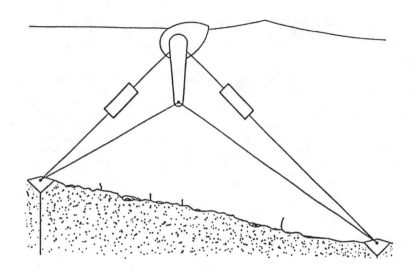

**Figure 2.  Solo Salter Duck on Tension Leg Mooring.**
**(Salter 1989)**

The moving wave collapses the bag. This is then reflated as the
wave falls. Air is pumped through the turbine on both strokes.

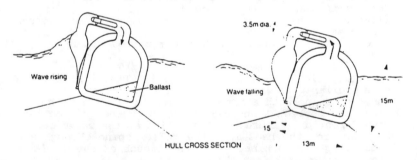

**Figure 3.  The Operation of the SEA Clam. (Davies ed. 1985)**

into a circular form is more effective than the long straight
spine.

OWC work at Queens University Belfast. Since 1987 a wave
energy research group headed by Whittaker (1990) has been
supervising the construction of Britain's first onshore
demonstration prototype OWC device, located at a natural gully
on the shore of the Isle of Islay, off the west coast of
Scotland. The civil engineering work was completed in
November 1988, followed by more than a year of testing, where
the pneumatic energy output was monitored. A biplane Wells
turbine was installed in 1990. The turbine has an induction
generator of 75 kW and flywheels having 2MJ inertial energy
storage, in order to smooth the effect of the variable
pneumatic energy input to the air turbine, Figure 4. The
system was connected to the Island's electrical grid in 1991.

Japanese Wave Energy Conversion Devices

        There has been a substantial amount of wave energy
research conducted in Japan, with many sea trials (Duckers and
Hotta 1990). A pioneer in the wave power research is Yoshio
Masuda who did his first sea tests in 1947, and is still
active in wave power, although he has now retired. In recent
years he has conducted full scale and model tests on a series
of wave energy converters, including the BBDB (backward bent
duct buoy) device, and a shore based design, similar to the
Norwegian Kvaerner Multiresonant unit (described later).
Masuda has been involved for many years in the development of
wave-powered navigation lights, he directed the KAIMEI project
(discussed in the next paragraph), and pioneered the use of
air turbines to generate electricity from wave energy.

        The perhaps most famous of the Japanese WEC work was
directed towards large scale testing at sea, on a barge
platform, KAIMEI ((Masuda and Miyazaki 1980). The platform
was 80 m (262 ft) long, 12 m (39 ft) beam, and 5 m (16 ft)
deep; it displaced 500 tons. The experiment was named after
the barge, and the work has been supported by the IEA (International Energy Agency). The early experiments have been
described as disappointing. The problem is believed to be
associated with the relatively short length of the barge (80
m), which results in excessive motions of the platform. The
overall conversion efficiency, based on the 80 m (262 ft)
length is approximately 4% (Masuda 1986). The barge is moored
parallel to the mean wave direction, and is thus an attenuat-
or. It has a number of oscillating water chambers in its
hull. The action of the ocean waves produces oscillations of
the water column in the barge; these movements generate
pneumatic power, which in turn is converted to electrical
power by means of air turbo-generators. The experimental
facility has been used to test moorings and several designs of
air turbines, air valves, and electric generators. Through
the collaboration of IEA, Japan was joined by four countries
(Canada, Ireland, UK and USA) in "Phase I"(1978-1980) and five

countries (Ireland, Norway, Sweden, UK and USA) in "Phase II" (1984-1987). The work on KAIMEI, completed in 1987, was conducted by JAMSTEC (Japan Marine Science and Technology Center), which continues to support wave energy work on new projects.

Several different Japanese devices have been tested at prototype scale, while others have been tested as models only. Several recent Japanese projects are described below.

Fixed Oscillating Water Column (OWC) Systems. A shore-based OWC prototype was built near Sanzei, on the west coast of Japan, by JAMSTEC, and a 40 kW tandem Wells turbine was tested there during the period from September 1983 to March 1984.

An OWC system, built into a breakwater has been developed for the Japanese Ministry of Transport (Goda et al, 1990). A 60 kW tandem Wells turbine/generator was installed in the machinery room, and electric power generation began in late 1989.

A 30 kW demonstration plant has been operating at Kujukuri, on Japan's Pacific Ocean coast, since March 1988 (Chino et al, 1989). The compressed air from the OWC unit drives a radial-flow turbine/generator, which powers a seawater pump for a flatfish aquaculture pond.

Ryokuseisha Corporation's Backward Bent Duct Buoy (BBDB). Masuda has reported that tests on scale models have demonstrated that this rather unusual configuration of floating platform (Figure 5) has superior performance to his earlier designs of floating wave energy devices, namely his navigation buoys and the barge KAIMEI (Masuda 1986). The buoy absorbs energy in heave and pitch and is claimed to have a performance which is two to three times better than the pneumatic navigation buoy and about ten times better than KAIMEI. The main components of the design are a long, horizontal, water-filled duct, with an opening facing in the direction opposite to the incoming waves, connected to a vertical chamber. This chamber has an air/water interface and an air turbine placed at the top, to develop power from the oscillating air flow. Design studies have shown that the length of the buoy (the horizontal duct) should be about 20-30% of the wavelength corresponding to the peak wave period at the proposed site. Model tests indicate that the buoy can convert to air power, about 59% of the incident wave power (59% "absorption width"). The conversion efficiency from air power to electricity is predicted to be 60%, so that the estimated "capture width", which is the product of these two efficiency values, is 35%. A small model (2.4 meter length) has been tested at sea as a navigation buoy in the Japanese sea and designs have been presented for a larger model, 24 meters length, 36 meters width, for application at various sites (Masuda 1990). For a site in Hawaii with 26 kW/m average wave power, the estimated

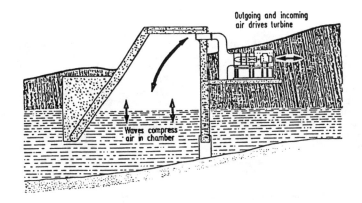

Figure 4. The Proposed Queen's University
Wave Energy Converter. (Salter 1989)

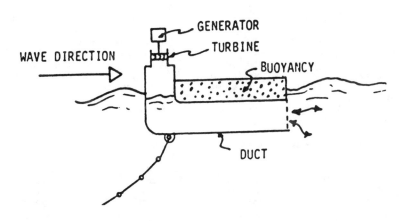

Figure 5. Backward Bent Duct Buoy (Masuda 1986)

average output for this wave energy converter is approximately 330 kW.

Pendulor. Developed at the Muroran Institute of Technology, the Pendulor is a pendulum gate which swings backwards and forwards in the opening at the front of a chamber facing the incident waves. The back of the chamber is a fixed wall, located a quarter wavelength (of the corresponding mean wave period) behind the axis of the pendulum pivot. The length of the pendulum gate is chosen in order to obtain resonance with the waves. The Pendulor's motion is damped by a back-to-back hydraulic system, which extracts energy from both halves of the oscillation. A 5 kW system has been in operation in the bay of Uchiura near to Muroran on Hokkaido. A similar unit to heat water is in operation on the west coast of Hokkaido, and three other units of 80 kW have been planned and designed.

A somewhat similar device called "Flap", differs from the Pendulor in that the hinge axis is submerged at the bottom of the flap. An experimental unit of 1 kW was constructed in 1989 and tested in the bay of Wakasa on the north west coast opposite Nagoya.

KAIYO. This device is a floating terminator employing hydraulic oil motors and electric generators. Wave energy is converted to hydraulic pressure energy through the relative motion between floating bodies and a stationary structure. The fixed structure, installed in 1984 at Iriomote Island, Okinawa, has a length of 26 m, breath of 20 m, height of 5.5 m and draft of 3.25 m, and it is located in 10 m of water. This part of the device is fixed to the sea bed by four 30 m long vertical legs, which can be jacked up and down to set the structure in the water or raise it clear. Two ship shaped floating sections, each 7.25 m long, 6.0 m wide and 4 m high with a draft of 2.25 m and a displacement of 68 $m^3$ are attached to the structure by links. A value of 13% of the length of this device has been given for the wave energy capture width (Duckers and Hotta, 1990).

Scandinavian Wave Energy Developments

An active program of experimental and theoretical research and development has been conducted, since the early 1970-s at various organizations in Norway. Analysis of point absorbers by Budal and Falnes (1977,1980,1982) introduced the concept of phase control in wave energy devices, to simulate resonance. This topic is discussed later in this chapter. Research and development was also conducted in Norway on various methods of wave focussing. Facilities were built to investigate various forms of focussing.

Since 1985 ocean tests have been conducted on the Norwegian west coast, 40 km north west of Bergen, with two different prototypes of wave energy converters, described below.

In Sweden an active development has been conducted by a research group in Gothenburg, since 1976. (Claeson et al.1987). One of the devices they have studied experimentally is the hosepump WEC which is described below. Since 1979 Kim Nielsen in Denmark has pursued development of the KN-device which is also described below.

The TAPCHAN Wave Power Plant. This Norwegian device converts the wave energy into potential energy in a reservoir on the shore. The principle is shown in Figure 6. The horn shaped collector is designed to collect the wave energy over a range of incoming wave frequencies and directions (Mehlum 1985 and Fredriksen 1986). The collector leads into the tapered channel with wall height of 10 m (from -7 m to +3 m). The waves enter the wide end of the channel from the collector and propagate down the narrowing channel with increasing wave height. The wave height is amplified until the crests spill over the walls into the reservoir at a level of 3 m above the mean sea level. In this way much of the wave energy is transformed into potential energy. The electricity is generated in a low head hydroelectric system designed for seawater. An induction generator of capacity 350 kW delivers electricity to the Norwegian grid. On average, 50 to 60 percent of the incident wave energy hitting the 40 m wide collector is converted to potential energy in the water reservoir. The demonstration prototype has been running since 1985, and is at present under consideration for a new demonstration prototype in Indonesia, and for a commercial WEC at King Island, Tasmania, as an alternative to a new diesel power station. For the commercial unit, the capacity will be between 1 MW and 1.5 MW, corresponding to the power demand on King Island (Ljunggren 1990).

The Kvaerner Multiresonant OWC. This is another Norwegian WEC, and uses a "resonant harbor" ahead of the pneumatic chamber to broaden the bandwidth of the absorber, Figure 7. The harbor has parallel walls, and this feature enhances the resonant characteristics of the oscillating water column (OWC) of the pneumatic converter. The medium-scale prototype has a width of 10m (32.8 ft) and was designed to have a maximum output of 500 kW, using a Wells turbine (Malmo and Reitan 1985). The ocean demonstration unit was installed in a steep cliff at a Norwegian island. The system converts wave energy very efficiently into pneumatic power. In terms of pneumatic power the capture width of this device is greater than the physical width of the device in many sea conditions. This confirms the theoretical analyses and the experimental results from small scale experiments. Due to the limited capacity of the machinery a control valve is installed in order to avoid turbine stalling. The electric generator is coupled to the local utility grid via an AC-DC-AC converter. Evidently, there is some power loss in each power converting component, with the result that the average capture width, in terms of electricity delivered to the grid is about half of the

Figure 6.   The TAPCHAN Wave Power Plant. (Fredriksen 1986)

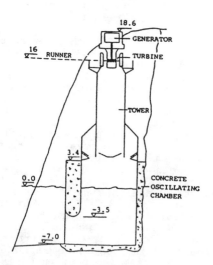

Figure 7.   The Kvaerner Multiresonant Oscillating Water
            Column Device.  (Source: Kvaerner Brug)

physical width of the WEC. Information of great value for the
design of an improved version of the WEC had been collected
when the steel tower of the device capsized in a heavy storm
in 1988. There was a plan to build a wave power station
forTonga in the South Pacific, but in connection with a
reorganization of the Kvaerner group of companies in 1989, it
was decided to curtail Kvaerner's involvement in wave power.

Götaverken's Hose Pump WEC. This project which has been conducted since 1980, is based on the "hose pump", a hose with
helical pattern of steel wire reinforcement, which causes the
hose to constrict when stretched, whereby the volume is
reduced (Claeson et al. 1987, p. 188). The hose is arranged
vertically below a heaving buoy, and the lower end of the hose
is fixed to a damping plate or to an anchor on the sea bed,
Figure 8.

          The hose is stretched and released as the buoy moves
up and down in waves. On the release, seawater is let in
through an inlet check valve, and as the hose is stretched,
seawater is pumped from the hose through an outlet check valve
into a collecting line. The flowing seawater from a group of
hose pumps enters the nozzle of a Pelton turbine housed in an
underwater chamber. A 64 MWe WEC consisting of 60 hose pumps
and 60 buoys with a common Pelton turbine has been designed
and analyzed. Each hose has 0.5 m inner diameter and 25 to 35
m length. The diameter of the buoys is 15 m and of the
damping plates 10 to 15 m. Sea tests have been performed,
using smaller models, off the Swedish west coast, 600 m from
the shore. Three hose pumps of internal diameter 0.2 m and
length of 10 to 15 m supplied pressurized sea water to a 30
kWe Pelton turbine and generator located on shore.

The KN Wave Power Converter. In Denmark Kim Nielsen has
conducted wave power research since 1979. After several storm
related disasters, an installation for a sea test has been
placed in the North Sea at a depth of 30 m, 3 km off Hanstholm
harbor. The turbine and generator are rated at 45 kWe. The
operating principle of this WEC is similar to the Götaverken
hose pump WEC. The main difference is that a piston pump on
the sea bed is used instead of a hose pump. A cable connects
the piston to a float on the sea surface, Figure 9. The
heaving float operating the piston pump has a diameter of 6 m
(Nielsen and Scholten 1989).

U.S. Wave Energy Developments

          The U.S. developments in wave energy have mainly
been funded from private sources, with some funding from the
Government. The U.S. Department of Energy has supported small
projects proposed by U.S. companies and has also been cooper-
ating with foreign governments in the research and development
activities on board KAIMEI.

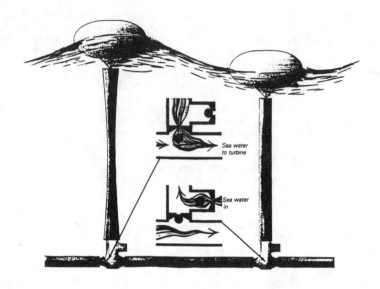

**Figure 8.    The Götaverken Hose Pump Concept. (Svensson 1988)**

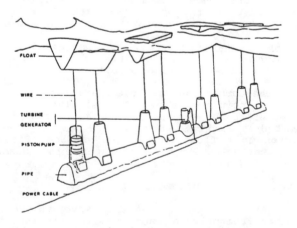

**Figure 9.    The Nielsen Wave Power Converter. (Nielsen 1986)**

Sea Energy Corporation, Articulated-Raft. This wave energy conversion device was developed in the U.S. with private support. The converter has one or more barge-like hulls hinged in line, Figure 10. The configuration is similar to an early British wave energy device, termed the Cockerell raft, and was patented about the same time. The energy is absorbed by hydraulic pumps near the hinge mechanisms, which respond to relative motions of adjacent barges. The Sea Energy Corporation has conducted model tests on large, 8m (26 ft) long models and has also completed engineering design work on their concept, as part of their technology development. The performance and cost estimates have been presented in a company report, and some technical information appears in a published paper (Burdette and Gordon, 1986).

Q Corporation, Tandem-Flap Wave Power Device. The Q Corporation has sponsored experimental and theoretical studies of rigid flap wave energy devices. The two flaps in the Tandem Flap device are hinged at the bottom edge and allowed to move in the direction of wave propagation. The motions of the flaps are restrained by linear springs and hydraulic dampers (power absorbers). The complete device is supported by a floating or fixed frame, Figure 11. Recent experimental studies have demonstrated good performance in regular waves. The company was expecting to conduct tests, using cost sharing funds from the Department of Energy, on an engineering model in the open water of the Great Lakes (Wilke, 1986).

Portuguese Wave Energy Program

The research and design work in Portugal on wave energy has been conducted at the Instituto Superior Tecnico, the Department of Renewable Energies, and the Department of Hydraulics, (Falcão 1990). Studies of the wave energy resource were used to evaluate various sites off the Portuguese coast, and off the islands of Madeira and the Azores. The selected location was on the northern coast of Pico Island in the Azores, where the average wave power is about 20 kW/m. A 1/35 scale model of the power plant and the surrounding region was built and tested in 1989. The plan is to build a 500 kW OWC unit off Pico Island, to be operational in 1992. The proposed design consists of a pneumatic chamber of rectangular plan form with two protruding side walls, similar to the Kvaerner Multiresonant OWC. The generator would be driven by a Wells turbine, perhaps modified to have controlled rotor blade settings to provide improved performance with phase control (Gato and Falcão, 1988).

Power Conversion Technology.

Wave energy is an irregular and oscillating low-frequency energy source while the electric grid requires a steady 60 Hz (or 50 Hz) supply. The power conversion device must take the energy supplied by the waves and convert it to the requirements of the grid network. Several methods have been

Figure 10.  The Sea Energy Corporation Single Raft Unit.
(Sea Energy Corporation, New Orleans, Louisiana)

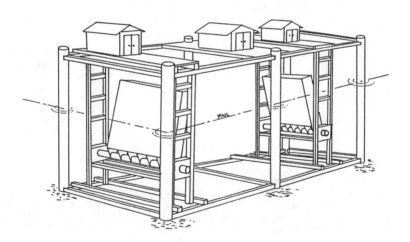

Figure 11.  The Fixed Q Corporation Tandem Flap Converter.
(Q Corporation 1987)

developed to do this and the most acceptable are pneumatic and hydraulic converters, because these systems are capable of transforming the energy from low frequency oscillations to high speed rotation, to drive conventional generators.

Pneumatic Systems

In the pneumatic systems an oscillating flow of low pressure air is produced by the motion of flexible bags (SEA Clam) or water air interfaces (KAIMEI, Kvaerner). The oscillating flow should be rectified and passed through an air turbine. The rectification can be provided by check valves, as in the early Masuda navigation buoy, or by self-rectifying turbines. With the check valves, the turbines can be of conventional design, but would be required to operate with interruptions to the flow as the air direction is changed.

Several designs of self-rectifying air turbines have been devised for wave energy systems. The simplest is the Wells turbine (Raghunathan 1980), which has a series of uncambered blades placed tangentially on the turbine disc as shown in Figure 12. This turbine operates at high rotational speed so that it is suitable for the low pressure levels in the various pneumatic wave energy devices. The Wells turbine is not as efficient as comparable conventional turbines, but is very simple to construct and, in many situations, requires neither guide blades, nor valves. The turbine has been used in the Kvaerner device, and also been tested on KAIMEI. Tandem and biplane configurations of the Wells turbine have also been evaluated.

Gato and Falcão (1988) have analyzed both theoretically and experimentally the performance of modified Wells turbines having blades with adjustable setting angles. The intention was to provide a method for improving the energy conversion from waves by phase control, using variable setting angles in the turbine for this purpose.

Another design of self-rectifying turbine was constructed to be tested on KAIMEI, the McCormick turbine. This is a counter-rotating turbine with a generator attached to each of the two rotors. Other self-rectifying turbines have been designed, and several patents have been issued for such devices (McCormick 1981).

The use of closed circuit pneumatic systems, as in the SEA Clam, would allow the air within the device to be separated from the corrosive environment of the seawater mists. Pneumatic turbines, either with rectifying valves or of self-rectifying design appear to be well developed for small power outputs, and should provide acceptable technology for wave energy conversion.

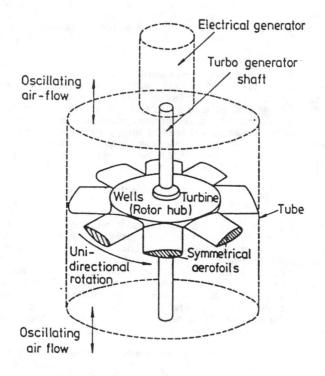

**Figure 12.   The Wells Self-Rectifying Air Turbine.**
**(Raghunathan 1980)**

Hydraulic Systems

        High pressure hydraulic systems would be used with
the Edinburgh Duck, Götaverken's Hose Pump, Sea Energy Corp.
Raft, Q Corp. Tandem Plate Converter, and other devices.
Seawater has been proposed as the hydraulic fluid for the hose
pump, while hydraulic oil would be utilized in most of the
other devices.   In all these systems the low-frequency
oscillating hydraulic flow is rectified and passed through a
turbine or motor, which drives a generator.   Energy storage
can be accomplished with an accumulator, reservoir, or
flywheel.   The technology for such systems is well established.

Phase Control

Phase control is the term used to describe a method for improving the efficiency of wave energy conversion devices at off-resonant conditions. The principles were described by Budal and Falnes (1982), and applied to a floating cylindrical buoy. It was recognized that the relative phase difference between the motion of the device and the incident wave force, had an important influence on the efficiency of conversion devices. By controlling this phase difference, Budal and Falnes had been able to improve the efficiency of a wave energy conversion device at frequencies below the natural resonant conditions of the device. In the experiments conducted by them, a small model buoy was mechanically stopped at various times during the motion, in order to provide the approximate motion of a much larger buoy at resonance. Phase control has also been considered and tested for OWC devices (Hotta et al. 1984, Perdigao and Sarmento 1989).

The apparent advantages of phase control are the possible reduction in size of wave energy conversion devices and the potential for broadening the range of frequency for efficient operation.

Force Reference

In practical designs of wave energy conversion devices it has been observed that it is difficult to provide a stable platform to support the device and to provide a frame of reference from which power can be generated. Floating platforms which respond to the waves, may reduce the output of the wave energy conversion devices supported by them. Bottom supported devices provide one solution. For floating devices, engineers have proposed tension leg moorings, long spines supporting many devices and spanning several waves, and the inertia and drag or other compensating forces of large underwater masses, to solve the problem.

Moorings and Cables

All types of floating devices in the ocean present problems with moorings and connecting power cables. The storm loads and the continual movement of the platform in waves and currents cause rapid wear, fatigue, and entanglement of the moorings and (riser) cables. Masuda (1986), described such problems with KAIMEI. The riser cables broke on several occasions and became tangled with the moorings in a storm. Masuda was apparently able to overcome these difficulties and developed a mooring and cable system that he stated could operate successfully for about two years before replacement.

Conclusions

Many different wave energy conversion devices have been designed and tested in the laboratory. New proposals are continuously being considered at various research centers throughout the world, and a few full sized devices, most of them shore-based, have been tested in the sea. With the present state of the art, wave energy development is still in its infancy. There is a need for further research and development. The emphasis should be on practical converters that are both efficient and able to withstand the rigors of the ocean environment.

## References

Bellamy, N.W., 1985. "The Circular SEA Clam Wave Energy Converter," in Evans, D.V. and Falcão, A.F.O. (Eds), 1985. IUTAM Symposium Lisbon 1985, *Hydrodynamics of Ocean Wave Energy Utilization*, (Springer-Verlag, Berlin, Heidelberg), pp. 69-79.

Berge, H.(Ed), 1982. *Second International Symposium on Wave Energy Utilization*, Proceedings, (Tapir, Trondheim, Norway, 1982).

Budal, K., 1977. "Theory for absorption of wave power by a system of interacting bodies," *Journal of Ship Research*, 21 (1977), pp. 248-253.

Budal, K. and Falnes, J.,1975. "A resonant point absorber of ocean-wave power," *Nature* 256 (1975), pp. 478-479.

Budal, K. and Falnes, J.,1977. "Optimum operation of improved wave-power converter," *Marine Science Communications* 3 (1977), pp. 133-150.

Budal, K. and Falnes, J., 1980. "Interacting point absorbers with controlled motion," in Count, B. (Ed), *Power from Sea Waves*, (Academic Press, London, 1980) pp. 381-399.

Budal, K. and Falnes, J., 1982. "Wave Power Conversion by Point Absorbers - A Norwegian Project," *International Journal of Ambient Energy*, Vol 3, April 1982, pp. 59-67.

Burdette, E.L. and Gordon, C.K., 1986. "Available Ocean Wave Power and Prediction of Power Extracted by a Contouring Raft Conversion System," Proceedings of ASME, *Offshore Mechanics Arctic Engineering Symposium*, Tokyo, 1986.

Chino, H., et al., 1989, "Verification Test of a Wave Power Generating System with a Constant Air Pressure Tank," *Air Conditioning with Heat Pump* (Department of Business and Economic Development, Energy Division, Honolulu, HI, copy on file) pp. 1-26.

Claeson, L. et al., 1987. "Energi från havets vågor," ("Energy from the ocean's waves" in Swedish), Efn-rapport nr. 21, *Energiforskningsnämnden*, Stockholm, 1987.

Count, B.(Ed) 1980a, *Power from Sea Waves*, (Academic Press,

London) 1980.

Count, B. 1980b, "Wave Power - A Problem Searching for a Solution," Count, B.(Ed), *Power from Sea Waves*, (Academic Press, London) 1980.

Davies, P.G. (Ed) et al.,1985. *Wave Energy*, The Department of Energy's R and D Programme 1974-1983 U.K. Department of Energy Report, ETSU R26, March 1985.

Duckers, L.J. (Ed), 1990. *Wave Energy Devices*, Conference Proceedings (C57). Solar Energy Society, London, March 1990.

Duckers, L.J. and Hotta, H., 1989. "A Review of Japanese developments," Duckers, L.J. (Ed) *Wave Energy Devices Conference Proceedings* (C57), London, March 1990, pp. 3-21.

Evans, D.V., 1976. "A theory for wave-power absorption by oscillating bodies," *Journal of Fluid Mechanics*, 77 (1976): 1, pp. 1-25.

Evans, D.V., 1979. "Some theoretical aspects of three-dimensional wave-energy absorbers," *Proceedings* of the Symposium on Ocean Wave Energy Utilization, Gothenburgh, Sweden, 1979, pp. 77-113.

Evans, D.V., 1980, "Some Analytic Results for Two and Three Dimensional Wave-Energy Absorbers," *Power from Sea Waves*, based on the proceedings of the conference on Power from Sea Waves, Edinburgh, 1979, pp. 213-249, (Academic Press London).

Evans, D.V., 1982, "A Comparison of the Relative Hydrodynamic Efficiencies of Attenuator and Terminator Wave Energy Devices," Berge, H.(Ed), 1982. *Second International Symposium on Wave Energy Utilization*, Proceedings, pp. 137-154, (Tapir, Trondheim, Norway, 1982).

Evans, D.V. and Falcão, A.F.O. (Eds), 1985. IUTAM Symposium Lisbon 1985, *Hydrodynamics of Ocean Wave Energy Utilization*, (Springer-Verlag, Berlin, Heidelberg).

Falcão, A.F. de O., et al, 1990. "Demonstration Wave Power Plant in the Azores: Preliminary Studies," World Renewable Energy Congress, Reading U.K., 1990.

Falnes, J., 1980. "Radiation impedance matrix and optimum power absorption for interacting oscillators in surface waves," *Applied Ocean Research*, Vol. 2, No 2, 1980, pp. 75-80.

Fredriksen, A.E., 1986, "Tapered Channel Wave Power Plants," in Twidell, J. and Lewis, C. (Eds) *Energy for Rural and Island Communities IV* (Pergamon Press Oxford U.K.) pp.179-182.

Goda, Y., et al, 1990. "Field Verification Experiment of a Wave Power Extracting Caisson Breakwater," in Krock, H.J.,

(Ed), *Proceedings* of the International Conference on Ocean Energy Recovery, (American Society of Civil Engineers, New York), pp. 35-42.

Gato, L.M.C. and Falcão, A.F.O., 1988. "Aerodynamics of the Wells Turbine: Control by Swinging Rotor-Blades," *International Journal of Mechanical Sciences*, Vol. 31, No. 6, pp. 383-395.

Hagerman, G.M. Jr., 1985. "Oceanographic Criteria and Site Selection for Ocean Wave Energy Conversion," in Evans, D.V. and Falcão, A.F.O. (Eds), 1985, IUTAM Symposium Lisbon 1985, *Hydrodynamics of Ocean Wave Energy Utilization*, (Springer Verlag, Berlin, Heidelberg), pp. 169-178.

Hagerman, G.M. and Heller, T., 1988. "Wave energy: A survey of twelve near-term technologies," *Proceedings* of the International Renewable Energy Conference, Honolulu, Hawaii, 18-24 September 1988, pp. 98-110.

Hotta, T., Miyazaki, T., Washio, Y. and Aoki, Y., 1984. "Fundamental study of phase control for air flow to improve on wave power absorption," *Journal of the Kansai Society of Naval Architects*, No 194.

Leishman, J.M. and Scobie, G., 1976. "The development of wave power. A techno-economic study," National Engineering Laboratory, East Kilbride, Scotland, 1976.

Lewis, T., 1985. "Wave Energy Evaluation for C.E.C.," (Graham & Trotman Ltd., London, 1985).

Li, G. and Guo, J., 1985. "Progress in China's Developmental Research on Wave Energy," Evans, D.V. and Falcão, A.F.O. (Eds), 1985. IUTAM Symposium Lisbon 1985. *Hydrodynamics of Ocean Wave Energy Utilization*, (Springer-Verlag, Berlin, Heidelberg), pp. 125-132.

Liang, X., Gao, X., Yu, Z., Jiang, N., Hou, X., and You, Y., 1991. "Research and Construction of Small Experiment Wave Power Station," Paper G-4, Third Symposium on Ocean Wave Energy Utilization, JAMSTEC Tokyo, January 22-23, 1991.

Ljunggren, S., 1990. Private communication.

McCormick, M.E., 1981. *Ocean Wave Energy Conversion*, (John Wiley & Sons, New York, 1981).

Malmo, O. and Reitan, A., 1985. "Development of the Kvaerner Multiresonant OWC." Evans, D.V. and Falcão, A.F.O. (Eds), 1985. IUTAM Symposium Lisbon 1985. *Hydrodynamics of Ocean Wave Energy Utilization*, (Springer-Verlag, Berlin, Heidelberg), pp. 57-67.

Masuda, Y., 1986. "An Experience of Pneumatic Wave Energy Conversion Through Tests and Improvement," McCormick, M.E., and Kim, Y.C. (Eds) 1987. American Society of Civil Engineers Symposium, La Jolla 1986. *Utilization of Ocean Waves,* (American Society of Civil Engineers, New York) pp. 1-33.

Masuda, Y. and Miyazaki, T., 1980. "The Sea Trial of the Wave Power Generator Kaimei," *Energy Developments in Japan,* Vol 3, No.2, October 1980, pp. 165-179.

Masuda, Y., 1990. "The Progress of Wave Power Generation in Japan," Presentation to PACOM, Souel, South Korea, June 1990.

Mehlum, E., 1985. "TAPCHAN." in Evans, D.V. and Falcão, A.F.O. (Eds), 1985. IUTAM Symposium Lisbon 1985, *Hydrodynamics of Ocean Wave Energy Utilization,* (Springer-Verlag, Berlin, Heidelberg), pp. 51-55.

Mei, C.C., 1976. "Power Extraction from Water Waves," *Journal of Ship Research,* vol. 20, No. 2, 1976, pp. 63-66.

Mei, C.C. and Newman, J.N., 1979. "Wave Power Extraction by Floating Bodies," *Proceedings* of 1st Symposium on Wave Energy Utilization, Gothenburg, Sweden, 1979.

Newman, J.N., 1976. "The interaction of stationary vessels with regular waves," *Proceedings* of 11th Symp. Naval Hydrodynamics, (1976) pp. 491-501.

Nielsen, K., 1986, "On the Performance of a Wave Power Converter," in McCormick, M.E., and Kim, Y.C. (Eds) 1987. American Society of Civil Engineers Symposium, La Jolla 1986. *Utilization of Ocean Waves,* (American Society of Civil Engineers, New York) pp. 164-183.

Nielsen, K. and Scholten, C., 1989. "Planning of a full scale wave power conversion test 1988-89," Paper at the International Conference on Ocean Energy Recovery, 28-30 Nov 1989.

Perdigao, J.N.B.A. and Sarmento, A.J.N.A., 1989. "A phase control strategy for OWC devices in irregular seas," J. Grue (Ed): The Fourth International Workshop on Water Waves and Floating Bodies. Dept. Mathematics, University of Oslo. pp. 205-209.

Q Corporation, 1987, "The Tandem-Flap Wave Power Conversion Device, The Great Lakes Experiment, Test Model Design/ Configuration Study, April 1987.

Raghunathan, S.R., 1980. "The Wells Turbine". The Queens University, Belfast, Department of Civil Engineering Report No. WE/80/11, 1980.

Raju, V.S., Ravindra, M., and Koola, P.M., 1991. "Energy from Sea Waves - the Indian Wave Energy Utilization Programme," Paper G-7, Third Symposium on Ocean Wave Energy Utilization,

JAMSTEC Tokyo, January 22-23, 1991.

Ross, D. and Townsley, M., 1989. "India picks up Britain's discarded wave project," *New Scientist*, 14 January 1989, p. 38.

Salter, S.H., 1974. "Wave Power," *Nature*, Vol 249, No.5459, 1974, pp. 720-724.

Salter, S.H., 1989. "World progress in wave energy - 1988," *International Journal of Ambient Energy*, Vol 10, Jan. 1989.

SEASUN Power Systems, 1988. *Wave Energy Resource and Technology Assessment for Coastal North Carolina,* North Carolina Alternative Energy Corporation, Research Triangle Park, NC, Contract UY 1873.

Shaw, R., 1982. *Wave Energy, A Design Challenge,* (Ellis Horwood Ltd., Chichester, England) 1982.

Stahl, A., 1892   "The Utilization of the Power of Ocean Waves," *Transactions* of the American Society of Mechanical Engineers, Vol 13, 1892.

Svensson, G. 1985. "Swedish Idea for Wave Power," *Energy Ahead 85*, Vattenfall, Annual Research and Development Report, Stockholm, Sweden.

Weigel, D.L., 1964. *Oceanographic Engineering*, (Prentice Hall, Englewood Cliffs, N.J.) 1964.

Whittaker, T.J.T., 1990. "Shoreline wave power on the Isle of Islay," Paper read 13 March 1990, to be published by the Institute of Engineers and Shipbuilders in Scotland.

Wilke, R.O., 1986. "Theoretical and Experimental Evaluation of an Engineering Model of the Tandem-Flap Wave Power Device," *Proceedings* of ASME, Offshore Mechanics Arctic Engineering Symposium, Tokyo, 1986.

Zossimov, Anatoly, 1990. Private communication.

# 9: ECONOMICS OF WAVE POWER

George Hagerman [1]

## Abstract

Although wave power has not yet moved into the commercial marketplace, it is much closer now than it was ten years ago. Within the last decade, several full-scale prototypes and demonstration plants have been built, ranging in size from 20 kWe to 500 kWe, and most of these are still operating. In the economic evaluation presented herein, levelized energy costs are computed for land-based, caisson-based, and offshore wave power plant designs derived from this recent experience.

Land-based Tapered Channel technology, developed in Norway, is on the verge of commercialization, with orders placed for two small plants in Australia and Indonesia. Japanese caisson-based systems are expected to enter the Pacific island market during the next five years. The development of offshore wave power is less advanced. Without long-term demonstration, private financing of an offshore project is unlikely, due to large uncertainties in absorber fabrication cost, plant maintenance requirements, component durability, and wave energy absorption efficiency.

For plant sizes of 30 MWe or less, the projected cost of wave-generated electricity is lower than that for other ocean energy technologies. At this scale, wave power is also less expensive than small hydropower or diesel generation. Where demand growth is sufficiently high to warrant construction of a coal-fired plant, however, wave power loses its competitiveness with fossil fuels, although it remains the least costly ocean energy option for all but the largest tidal power plants.

Several co-products enhance wave power's economic appeal, including seawater pumping for aquaculture, fresh water production by direct reverse osmosis, and breakwater protection. The case is made that for island applications, fresh water may be more valuable than electricity, in terms of both market size and energy impact.

Given a scenario of commercial success in island markets, long-term demonstration of offshore components and systems, and the development of an onshore infrastructure for the storage and distribution of hydrogen, wave power can ultimately supply a large fraction of global demand for electricity, heating, and transportation fuel.

---

[1] SEASUN Power Systems, 124 E. Rosemont Ave., Alexandria, VA 22301

## Introduction

As of this writing, there is less than half a megawatt of grid-connected wave generating capacity operating worldwide, distributed among five demonstration plants. These are not commercial projects, however, in the sense that their construction and operation is supported entirely by revenue from the sale of electricity. All have received financial assistance from national, regional, or local governments, as well as private-industry funding from within the developing companies.

This chapter is divided into six text sections. The first section describes the methodology used to estimate energy costs, and the next section presents results for land-based, caisson-based, and offshore power plant designs. Capital cost is the most important component of any wave energy economic assessment, and its breakdown is analyzed in the third section. The fourth section describes the economic risks likely to be encountered in the development of a wave power project and how they can be reduced. The fifth section reviews various wave energy applications and outlines alternative market development scenarios. The last section compares the life-cycle cost of wave power with that of other energy technologies, drawing on examples from the southwest Pacific Ocean, Hawaii, northern California, and the United Kingdom.

## Economic Assessment Methodology

Levelized energy costs were computed using financial parameters specified in the Electric Power Research Institute Technical Assessment Guide (EPRI TAG™; parameter values are given in the Appendix at the end of this chapter). These reflect the life-cycle costs that would be experienced by a regulated, investor-owned utility in the United States. It should be noted, however, that the first commercial wave power plants probably will be financed as independent projects, where the owner/operator sells power to a utility. There is no standard formula for computing energy costs in such cases, because of the wide variability in project financing schemes. For example, the plant may be sold to the utility after an initial period of successful operation under a power-purchase agreement, but well before the end of its service life. Under these circumstances, a single levelized energy cost would be difficult to define.

Nevertheless, a standard formula is required if economic comparisons are to reflect differences in technology cost-effectiveness rather than differences in accounting methods, financing schemes, or government monetary policies. The EPRI TAG™ was chosen because it has been developed for just this type of analysis and is well documented (Electric Power Research Institute, 1987).

Capital and operating cost estimates originating outside the United States were converted to U.S. dollars using the exchange rate in effect at the time the original estimate was made (see Appendix). All original-year costs were escalated to 1990 at an annual inflation rate of 6%.

Where capital cost breakdowns were available, absorber structure typically accounted for the largest share of the plant's direct cost. It was also evident that concrete and steel fabrication costs varied significantly from country to country. In order to more clearly reveal economic trends related to the efficient use of structure, a standard set of unit fabrication costs was developed. These costs are thought to be typical of small U.S. fabrication yards, and are given in the Appendix.

## Levelized Energy Costs

It is significant that the five demonstration plants now operating are all land- or caisson-based systems. Floating devices are less well-developed, and as a result, assessments of their economic feasibility are less certain.

On the other hand, the cost and performance of land- and caisson-based systems are more sensitive to local site conditions. The cost of site preparation for a Tapered Channel power plant or land-based oscillating water column depends greatly on shoreline geology and topography. To the extent that cast-in-place concrete construction is required, the economic feasibility of these schemes also depends on the local availability of aggregate material. Likewise, the cost of a wave energy breakwater is influenced by the need for seafloor foundation leveling, as well as the local availability of suitable rock for a rubble mound on which the caissons would rest.

The sheltering effect of coastal features, such as headlands and peninsulas, and the effects of wave refraction and shoaling are much greater nearshore than offshore. Consequently, the performance of land- or caisson-based systems for a given deep-water wave resource is highly dependent on the exact coastal location of the plant.

Therefore, while cost and performance projections for land- and caisson-based wave power plants draw on greater experience, they are also more site-specific. Projections for offshore systems are less certain, but are more generally applicable for a given regional wave climate.

## Land- and Caisson-Based Systems

As previously mentioned, less than half a megawatt of grid-connected wave generating capacity is operating worldwide. This is overwhelmingly dominated by a 350 kWe Tapered Channel plant at Toftestallen, on the North Sea coast of Norway, which has been operating continuously since the third quarter of 1986. Developed by Norwave A.S., this demonstration plant has survived several severe storms, including one which destroyed a nearby cliff-based oscillating water column system.

Norwave is the only company that can be said to have realistic prospects for commercial wave power development in the near-future. Based on the proven durability of the Toftestallen project, orders have been placed for two 1.5 MWe plants:

one on the south coast of Java, in Indonesia, and the other on King Island, in Bass Strait, just north of Tasmania.

Cost and performance projections for three Tapered Channel projects are given in Table 1. The following comments are based on telephone conversations with Even Mehlum (1990), a founder and principle owner of Norwave A.S.

The reservoir for the Toftestallen demonstration was built by damming two small inlets to the island's interior bay (Figure 1). A collector channel, 2-3 m wide and 7 m deep, was mined into the rock at the head of a natural gully. The gully is 60 m wide at its entrance and funnels waves into the collector channel, which leads to the converter. The concrete walls of the converter are formed to a height 3 m above sea level, and as waves travel along its 60 m length, they spill water over its sides and into the reservoir. Water then drains back to the sea through a Kaplan turbine/generator.

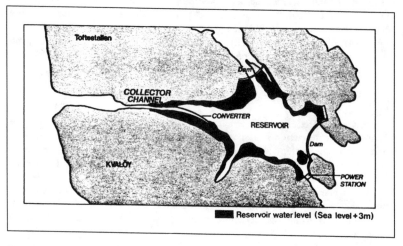

FIGURE 1. Tapered Channel demonstration plant. (Modified after JEFFREYS, 1985.)

The plant's annual output given in Table 1 is based on a projected wave energy resource of 25 kW/m, whereas the incident wave power measured at Toftestallen during the past five years has averaged only 7 kW/m. It should also be noted that the projected capital cost is based on an identical plant built at the same site, incorporating "lessons learned" during the construction of the demonstration plant and not requiring extensive monitoring instrumentation.

The purpose of the Toftestallen project was to verify wave/channel hydrodynamic performance rather than optimize electric power generation. In order to keep costs down, a 350 kWe generator was used, even though it is much smaller than optimal for this size reservoir in the projected wave climate.

TABLE 1. Cost and performance of Tapered Channel power plants. [a]

| Location: | Toftestallen (Norway) | Java (Indonesia) | King Island (Australia) |
|---|---|---|---|
| Average wave power: | 25 kW/m | 20-25 kW/m | 30-32 kW/m |
| Collector opening width: | 55 m | unknown | unknown |
| Reservoir surface area: | 5,500 m$^2$ | 8,100$^2$ | unknown |
| Plant capacity: | 0.35 MWe | 1.5 MWe | 1.5 MWe |
| Annual output: | 1.7 GWh | 10 GWh | 9 GWh |
| Capacity factor: | 25% | 76% | 68% |
| Initial investment: | $3,560/kWe | $2,000/kWe | $3,290-3,840/kWe |
| Annual O&M cost: | $18/kWe | $33/kWe | $33/kWe |
| Cost of energy: | 8.0 ¢/kWh | 3.6 ¢/kWh | 6.2-7.2 ¢/kWh |

[a] A plant service life of 40 years and a construction time of one year was used for all projects. Costs are in 1990 constant dollars. See Appendix for financial parameters and currency exchange rates. (Data from NORWEGIAN ROYAL MINISTRY OF PETROLEUM AND ENERGY, 1987; MEHLUM, 1990.)

The Java plant best represents the relationship between generator capacity, plant output, and reservoir size. Its capital cost is less than two-thirds that of the King Island project, even though both plants have the same generation capacity. This is largely due to the remoteness of the King Island site, which requires the construction of a 5 km access road and overland transmission line.

Given the more energetic wave climate at King Island, Norwave expects a higher capacity factor there, but the plant's annual output is limited by the small size of the island's utility grid. This emphasizes the importance of accurate demand forecasting; no matter how cost-effective a wave energy device is at a given site, the project won't be economically feasible if there isn't sufficient demand for the generated electricity.

Norwave projects a 40-year service life for its Tapered Channel power plants, which are quite similar to long-lived, conventional, hydroelectric plants. Construction of the 1.5 MWe projects is expected to take 12 to 18 months.

Turning next to caisson-based systems, Table 2 presents cost and performance data for three oscillating water column (OWC) designs, and one pivoting flap system. The Toftestallen and Tongatapu projects were based on stand-alone caissons. In the two Japanese designs, however, the caissons are placed next to one another, forming a continuous breakwater.

TABLE 2. Cost and performance of caisson-based wave power plants. [a]

| Device: | Double-Acting MOWC | Double-Acting MOWC | Single-Acting OWC | Pivoting Flap |
|---|---|---|---|---|
| Developer: | Kvaerner Brug A/S | Kvaerner Brug A/S | Takenaka Corporation | Muroran Institute of Technology |
| Location: | Toftestallen (Norway) | Tongatapu (Tonga) | Kashima Port (Japan) | Muroran Port (Japan) |
| Water depth: | seaside cliff | reef crest | 4.5 m | 5 m |
| Average $H_S$: | unknown | unknown | 1.4 m | 2.0 m |
| Average $T_S$: | unknown | unknown | 8 sec | 6 sec |
| Average wave power: | 15 kW/m | 21 kW/m | unknown | 11 kW/m |
| Caisson width: | 10 m | 18 m | 7 m | 25 m |
| Number of caissons: | one | four | 20 | 10 |
| Plant capacity: | 0.5 MWe | 2 MWe | 1.5 MWe | 1.5 MWe |
| Annual output: | 0.64 GWh | 5.8 GWh | 4.1 GWh | 6.9 GWh |
| Capacity factor: | 15% | 33% | 31% | 53% |
| Initial investment: | $1,610/kWe | $4,490/kWe | $7,320/kWe | $7,030/kWe |
| Re-investment: | none | none | none | $2,560/kWe every 15 yr |
| Annual O&M cost: | $16-32/kWe | $45-90/kWe | $57/kWe | $70/kWe |
| Plant service life: | 25 yr | 25 yr | 25 yr | 45 yr |
| Cost of energy: | 16-17 ¢/kWh | 19-21 ¢/kWh | 33 ¢/kWh | 20 ¢/kWh |

[a] A construction time of one year was used for all projects. Costs are in 1990 constant dollars. See Appendix for financial parameters and currency exchange rates. (Data from BONKE AND AMBLI, 1987; TOMMERBAKKE, 1988A; WATABE AND KONDO, 1989; CHINO, NISHIHARA, AND NAKAKUKI, 1989.)

The largest wave power plant built to date was developed by Kvaerner-Brug A/S, a large Norwegian hydropower company. Kvaerner Brug's caisson design consists of a rectangular capture chamber, and a harbor formed by extending the side walls of the chamber in a seaward direction. Hydrodynamic interaction between the harbor and capture chamber causes the OWC to have several natural frequencies, and this concept is referred to as a multi-resonant oscillating water column, or MOWC.

A 500 kWe demonstration plant based on this concept was built at Toftestallen, alongside Norwave's Tapered Channel project. The turbine was a monoplane Wells design. The resonant harbor was excavated out of the island's cliff wall (Figure 2).

FIGURE 2. MOWC demonstration plant at Toftestallen. (Courtesy of Kvaerner Brug A/S.)

From November 1985 until December 1986, electric power was generated only when personnel were present to run tests and conduct measurements. The plant began fully automatic operation in January 1987, running for two years until it was destroyed by a severe storm in January 1989 (Reuter News Service, 1989).

As with Norwave, Kvaerner Brug's Toftestallen cost estimate is based on an identical plant built at the same site rather than the actual cost of the demonstration plant itself. Also note that the 15 kW/m resource for Kvaerner Brug's performance estimate is based on the average incident wave power *when plant output was measured*, and so does not represent an annual average. While Norwave's turbine/generator was undersized for the reservoir at Toftestallen, Kvaerner Brug's 500 kWe unit was too large for a 10 m caisson, which accounts for its projected low capacity factor (Tommerbakke, 1988b).

Kvaerner Brug also developed a commercial plant design for the south Pacific Kingdom of Tonga. This project, now abandoned, better represents the relationship between turbine/generator capacity and capture chamber size. The plant's capacity factor more than doubles, but the larger capture chamber involves a considerably higher concrete fabrication cost. The resonant harbor would have been formed as part of the caisson, and the reef crest site at Tongatapu would have involved more excavation than was required at Toftestallen (Figure 3).

FIGURE 3. Caisson-based MOWC design developed by Kvaerner Brug for Tongatapu. Each caisson houses a 500 kWe Wells turbine/generator. (From STEEN, 1985.)

Kvaerner-Brug's MOWC is double-acting, in that the Wells turbine absorbs energy from both the rise and fall of the oscillating water column. A single-acting OWC device, which absorbs energy only from the water column's rise, has been developed by Takenaka Corporation in Japan. It consists of relatively narrow caissons which act as capture chambers. Output from several caissons is manifolded into a high-pressure air tank, which drives a conventional impulse turbine. Air is drawn into the caissons through check valves as the water column falls. When the water column rises, the air is directed to the manifold through a different set of check valves (Figure 4).

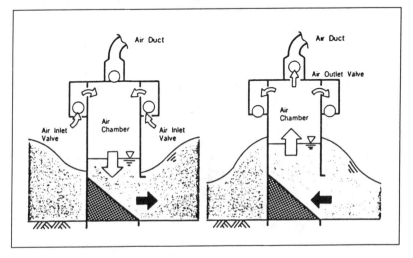

FIGURE 4.    Takenaka Corporation's single-acting OWC concept (Modified after CHINO, NISHIHARA, AND NAKAKUKI, 1989.)

A 30 kWe demonstration plant has been operating at Kujukuri, on Japan's Pacific Ocean coast, since March 1988 (Chino, Nishihara, and Nakakuki, 1989). Ten steel caissons, 2.1 m in diameter and 11.8 m high, are connected by an 0.8 m diameter duct to an onshore cylindrical air tank. The air tank floats in a ground tank of water and can be weighted with ballast, such that it sinks deeper into the ground tank, increasing the pressure against which the OWC units must work. The compressed air drives a radial flow turbine/generator, which powers a seawater pump for a nearby flatfish aquaculture pond.

Based on their operating experience with this plant, Takenaka Corporation has developed a conceptual design for a 1.5 MWe plant at Kashima Port (Figure 5). It should be noted that their capital cost estimate does not include fabrication of the wave-absorbing caissons, which is assumed to have been financed by some other source for shore or harbor protection.

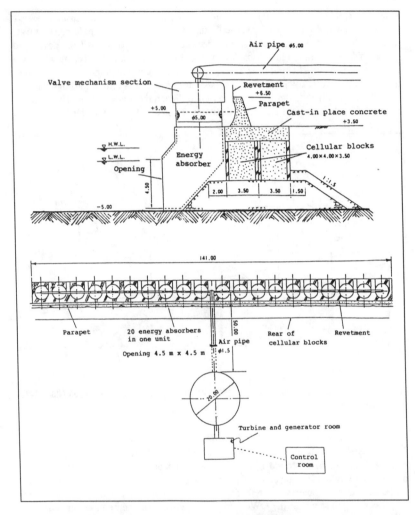

FIGURE 5. Conceptual design of a 1.5 MWe single-acting OWC plant, developed by Takenaka Corporation for Kashima Port. (From CHINO, NISHIHARA, AND NAKAKUKI, 1989.)

For over a decade, the Muroran Institute of Technology has developed a caisson-based pivoting flap device, which it calls the Pendulor System. The Pendulor itself is a stiffened steel plate that hangs down from the roof of a capture chamber. Unlike fixed OWC systems, where the capture chamber extends below the sea surface, the Pendulor System capture chamber is entirely open to the sea (Figure 6).

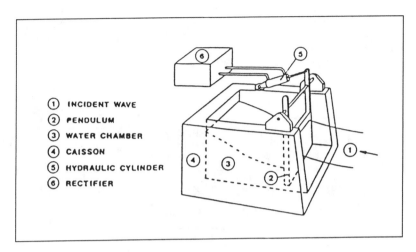

FIGURE 6. Caisson-based pivoting flap concept developed by Muroran Institute of Technology. (From KONDO AND WATABE, 1990.)

Incident waves interact with waves reflected off the back of the chamber to create a standing wave. The Pendulor is positioned at the node of this standing wave, where horizontal forces are at a maximum. As the Pendulor swings in response to these surge forces, it drives a double-acting hydraulic cylinder, which in turn drives a pair of hydraulic motors coupled to an electric generator.

In April 1983, a 5 kW (hydraulic motor rating) prototype was installed at Muroran Port, on the south coast of Hokkaido. The prototype caisson is sited in front of an existing seawall, in a water depth that ranges from 2.5 m at low tide to 4 m at high tide. Two capture chambers have been built into the caisson, but only one has been fitted with a Pendulor.

Twenty months after its installation, the Pendulor was bent during a severe storm, and the shock absorbers for the end-stops, which prevent over-stroking of the cylinder, had to be redesigned. A new Pendulor was installed in November 1985, and the prototype has survived several severe storms since then, without damage.

A small Pendulor System that generates electric power was deployed in 1981. Rated at 20 kWe, this unit is used to heat the public bath of a fishing cooperative at Mashike Harbor, on Hokkaido's west coast (Kuroi, 1983). Unfortunately, its Pendulor was also damaged by a storm, just three months after installation. It was replaced by a shorter Pendulor in 1983, which left a considerable gap at the bottom of the capture chamber. While this has prevented further damage, it has also lowered the system's conversion efficiency (Watabe, 1990). Nevertheless, the plant continues to operate.

Based on their experience with these two systems, Watabe and Kondo (1989) have developed a conceptual design for a 1.5 MWe plant at Muroran Port (Figure 7). Unlike Takenaka Corporation, their projected initial capital investment does include the caissons, and these account for 75% of the total plant cost.

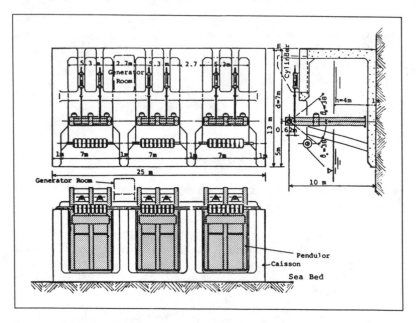

FIGURE 7. Caisson-based pivoting flap design. Each caisson contains three Pendulors, which drive a 150 kWe power conversion system. (Modified after KONDO AND WATABE, 1989.)

Watabe and Kondo (1989) project a 50-year service life for the breakwater caissons, but only a 15-year life for the balance of the plant. A 45-year plant life was used in Table 2, because it would not be cost-effective to replace balance-of-plant components at the end of the 45th year, for only five more years of caisson life.

Although incident wave power estimates were not published in the Takenaka design study, some inferences can be made by comparing wave statistics between Muroran Port and Kashima Port. Incident wave power is directly proportional to the square of significant wave height ($H_S$), multiplied by significant wave period ($T_S$). Thus, incident wave power at Kashima Port is about 35% less than that at Muroran Port, while Takenaka Corporation's proposed breakwater length is 45% shorter. If both plants were equally efficient, then the annual electricity output of Takenaka Corporation's design should be only about 2.5 GWh (65% x 55% x 6.9 GWh), but at 4.1 GWh, it is considerably higher.

This suggests that Takenaka Corporation's OWC is more efficient than Muroran Institute's pivoting flap. At $7,030/kWe, it is also much more expensive. Apart from the breakwater caissons, Muroran's balance-of-plant cost is only $2,560/kWe. If caisson fabrication and deployment can be financed by some other source (as has been assumed by Takenaka Corporation), and all other parameters in Table 2 remain the same, then the levelized energy cost at Muroran Port would be only 9.2 ¢/kWh over the 15-year service life of the balance-of-plant components.

In passing, it should be noted that annual O&M costs for the Kvaerner Brug designs were estimated by the developers to be 1-2% of initial investment. Muroran Institute of Technology uses a 1% estimate, whereas Takenaka Corporation has projected O&M costs on the more accurate basis of annual materials and labor requirements.

## Offshore Systems

There has been much less full-scale experience with floating wave energy devices than with land- or caisson-based systems. The specially built ship, *Kaimei*, was deployed twice in the Sea of Japan, to test various pneumatic turbines (used in OWC systems and the SEA Clam). Electricity was landed ashore from only a single 125 kWe turbine/generator, however, and this only for four months during the winter of 1978-79 (Ishii, et al., 1982). Two heaving buoy systems have also been grid-connected for a period of months: a 30 kWe Swedish system in 1983-84, and a 45 kWe Danish system in the spring of 1990.

Within the past five years, detailed economic feasibility studies are available for only three offshore devices that have been tested outside the laboratory wave tank:

- A 12.5 MWe circular-spine SEA Clam design, developed as part of the United Kingdom's wave energy program, for the Outer Hebrides Islands off Scotland's northwest coast (Bellamy, 1992).

- A 30 MWe heaving buoy design, based on the Swedish hose pump concept. The economic feasibility of this design has been studied for an offshore site at Half Moon Bay, just south of San Francisco, California (SEASUN Power Systems, 1991), and for a location off Makapuu Point, on the east coast of Oahu, Hawaii (SEASUN Power Systems, 1992).

- A 740 MWe central station design, based on the Danish heaving buoy system, developed by Kim Nielsen. This design was prepared for a North Sea site off the coast of Denmark (B. Hojlund Rasmussen, 1988).

None of the above designs have been considered for actual deployment. They originally were prepared simply to provide an indication of economic feasibility. Each design is described briefly and illustrated in the following pages.

The circular SEA Clam design consists of five 2.5 MWe modules. Each module has a toroidal hull, 60 m in diameter and 8 m deep, assembled from twelve concrete sections (Figure 8). Bags from adjacent sections communicate with each other via metal ductwork. Ten 250 kWe Wells turbine/generators are placed in-line with the ductwork, and the entire air system is maintained at a pressure of 15 kPa (2 psi).

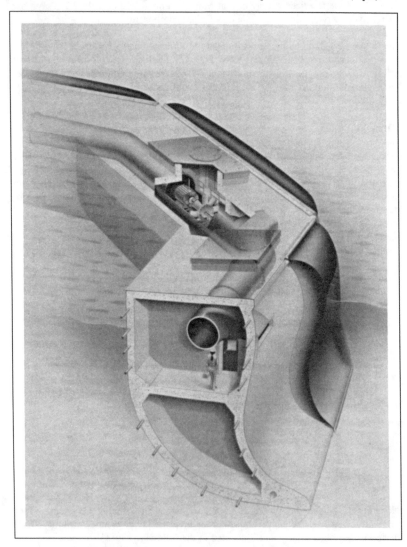

FIGURE 8. Circular SEA Clam design - concrete hull. (Courtesy of Sea Energy Associates, Ltd.)

The Swedish heaving buoy design consists of three star-shaped clusters of buoys, with six collecting lines per star, and ten buoys per collecting line (Figure 9). Each buoy has a diameter of 16-17 m. The buoys are spaced on 30 m centers, giving each star a total diameter of about 600 m. The underwater habitat at the center of each star houses a 10 MWe Pelton turbine/generator. The three stars are linked by a three-phase AC cable, which transmits power to shore at 35 kV.

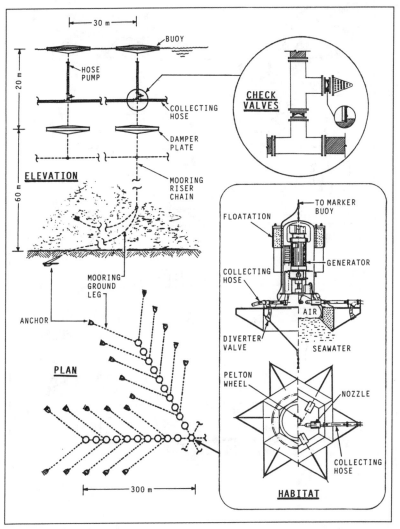

FIGURE 9. Swedish heaving buoy design. (Habitat detail courtesy of Technocean AB.)

In the Danish heaving buoy system, each converter consists of a 10 m diameter buoy tethered to a concrete piston. The piston is contained within a reinforced concrete cylinder, which is part of a circular deadweight anchor (Figure 10).

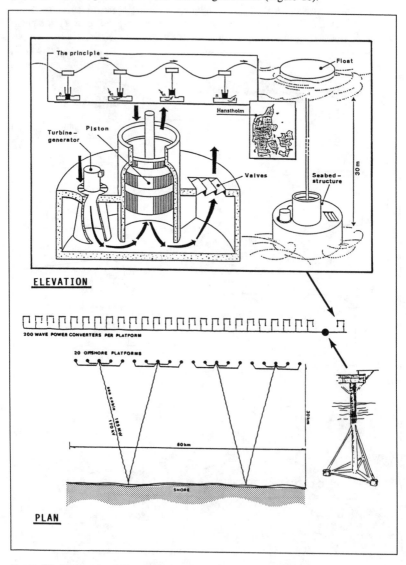

FIGURE 10.  Danish heaving buoy design.  (Elevation courtesy of Danish Wave Power, Aps; plan modified after B. HOJLUND RASMUSSEN, 1988.)

A 185 kWe Flygt submersible turbine/generator is mounted on the anchor, such that when the buoy heaves up on a wave crest, the piston draws water through the turbine into the anchor base. When the piston drops down in a wave trough, water returns to the sea through a one-way flap valve in the anchor wall. The electrical output from 200 modules is transmitted at 3 kV to a fixed platform, and the output of twenty such platforms is then transmitted to shore at 170 kV.

Cost and performance projections for the four different offshore designs are compiled in Table 3. An analysis of these results is presented following the table.

TABLE 3.  Cost and performance of offshore wave power plants. [a]

| Device: | Circular SEA Clam | Swedish Heaving Buoy | Swedish Heaving Buoy | Danish Heaving Buoy |
|---|---|---|---|---|
| Location: | Lewis (Scotland) | Half Moon Bay (California) | Makapuu Point (Oahu, Hawaii) | Hanstholm (Denmark) |
| Distance offshore: | 5 km | 24 km | 10 km | 30 km |
| Water depth: | 80 m | 80 m | 80 m | 30 m |
| Average $H_s$: | unknown | 2.0 m | 1.9 m | unknown |
| Average $T_p$: | unknown | 10.8 sec | 8.2 sec | unknown |
| Average wave power: | 52 kW/m | 25 kW/m | 15 kW/m | 9.1 kW/m |
| Absorber width: | 60 m | 17 m | 16 m | 10 m |
| Number of absorbers: | five | 180 | 180 | 4000 |
| Plant capacity: | 12.5 MWe | 30 MWe | 30 MWe | 740 MWe |
| Annual output: | 27.9 GWh | 76.3 GWh | 111 GWh | 870 GWh |
| Capacity factor: | 25% | 29% | 42% | 13% |
| Initial investment: | $2,210/kWe | $2,100/kWe | $1,970/kWe | $1,790/kWe |
| Re-investment: | none | $143/kWe | $173/kWe | none |
| Annual O&M cost: | $88/kWe | $84/kWe | $79/kWe | $10/kWe |
| Plant service life: | 25 yr | 30 yr | 30 yr | 20 yr |
| Construction time: | 1 yr | 1 yr | 1 yr | 5-10 yr |
| Cost of energy: | 15 ¢/kWh | 13 ¢/kWh | 8.6 ¢/kWh | 22-26 ¢/kWh |

[a]  Costs are in 1990 constant dollars. See Appendix for financial parameters and currency exchange rates. (Data from BELLAMY, 1992; SEASUN POWER SYSTEMS, 1991; SEASUN POWER SYSTEMS, 1992; B. HOJLUND RASMUSSEN, 1988.)

Comparison of the two Swedish heaving buoy designs reveals several interesting features. Oahu's island shelf is much narrower than the continental shelf off Half Moon Bay, and only a 10 km power cable is required to reach the 80 m depth contour. Smaller diameter buoys are optimal in the shorter period waves off Makapuu Point, which represents an even bigger cost saving. On the other hand, offshore deployment equipment is expected to be more costly in Hawaii, which results in a higher re-investment for periodic replacement of hose pumps, collecting lines, and mooring hardware. The initial capital investment, however, is less, as higher deployment costs are more than offset by lower power cable and buoy fabrication costs.

The 30 MWe reference design has a much greater capacity factor in Hawaii than off northern California. This is because the buoys are more efficient absorbers in shorter period waves, which are predominant along Hawaiian island coasts exposed to northeast trade winds. On the other hand, the wave energy resource off northern California is dominated by long-period swell, in which buoy performance is much poorer. Off Makapuu Point, the average wave energy absorption efficiency is 43%, compared with only 17% off Half Moon Bay. Trade wind waves are also more persistent than swell generated by winter storms in the north Pacific, which further accounts for the higher capacity factor in Hawaii.

Only the Swedish heaving buoy designs account for re-investment in capital equipment, yet both of the other systems will have periodic capital expenses. The submersible Flygt turbine/generators of the Danish heaving buoy system will require overhaul once every 3 to 5 years (Nielsen and Scholten, 1990). While the cost of overhaul parts may not be significant, the mobilization and deployment of offshore equipment to accomplish this work could be considerable. Similar considerations hold for periodic replacement of the SEA Clam's flexible bags, and if these cannot be changed at sea, then the Clam hull will require dry-docking, in addition to the offshore equipment spread for retrieving and re-deploying the unit.

The service lives listed in Table 3 are those projected by the developers. The Danish heaving buoy design, however, should be able to realize the same 25- to 30-year service life projected for other offshore systems. A short service life, coupled with a long build-out time, contribute significantly to the higher cost of energy for this design. The relatively poor wave energy resource off Hanstholm is also an important factor.

Annual operation and maintenance (O&M) costs for the Swedish heaving buoy designs were estimated as 4% of initial capital investment, based on offshore oil and gas industry experience. The annual O&M cost for the SEA Clam was developed from a line-item estimate, and works out, coincidentally, to be 4% of the capital cost.

Estimating O&M cost as a simple percentage of initial capital investment is not nearly as accurate as a line-item estimate. Furthermore, in the words of Stephen Salter (1988), "it predicts exactly the wrong outcome of spending money to reduce maintenance costs". Long-term ocean testing is required in order to truly understand the inspection, maintenance, and repair requirements of offshore wave power plants.

## Capital Cost Analysis

As with other renewable energy systems, the cost of wave-generated electricity is dominated by fixed charges associated with a high initial capital investment. Capital cost breakdowns are available for four of the systems described above, and these are presented in Figure 11. Each capital cost center is reviewed below.

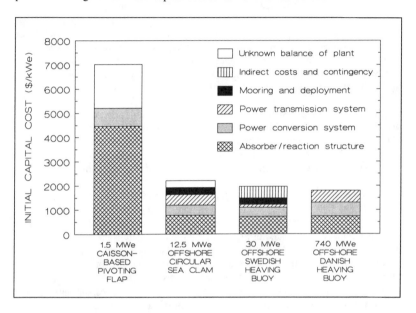

FIGURE 11. Capital cost breakdowns for four wave power plant designs. (Data from WATABE AND KONDO, 1989; BELLAMY, 1992; SEASUN POWER SYSTEMS, 1992; B. HOJLUND RASMUSSEN, 1988.)

In all designs, fabrication of absorber and reaction structures is the single biggest capital cost, ranging from 35% to 64% of the total investment. It should be noted that the original Danish cost estimate for buoys and gravity bases included the cost of deploying these structures. In this analysis, however, the absorber/reaction structure cost is based on a standard schedule of material fabrication costs, given in the Appendix. It was not possible to determine what part of the original Danish estimate was due to deployment, which is why it is missing from the breakdown in Figure 11.

The mechanical and electrical plant that converts absorber forces and motions into fluid power and thence to electric power, accounts for 19% to 32% of the capital cost of the three offshore systems, but only 10% for the caisson-based pivoting flap. This suggests that the power conversion system for this design may be undersized, as further evidenced by its high capacity factor in a relatively low-energy wave climate.

Sea-to-shore transmission costs are 20% or less for the smaller offshore systems. By comparison, the large central station design of the Danish heaving buoy system incorporates twenty fixed platforms as part of its transmission scheme, which accounts for 27% of the capital cost of that design.

Mooring and deployment account for 13% of the capital cost in the SEA Clam design and 14% in the Swedish heaving buoy system. As mentioned above, it was not possible to break out this cost center from the Danish heaving buoy design. Deployment costs were also not specifically accounted for in the caisson-based pivoting flap design. It should be noted that deployment of offshore wave power plants is likely to be more weather sensitive than it is for caisson-based systems. In the Swedish heaving buoy designs, a 30% "waiting-on-weather" contingency was applied to all buoy and habitat deployment activities.

Of the three offshore systems, only the Swedish heaving buoy design is known to account for indirect costs. These include land-based general facilities (office building, shop, warehouse, and associated utilities; does not include buoy fabrication yard, which is accounted for in absorber/reaction structure cost center), field engineering, and home office fees. Indirect costs were estimated to be 20% of the direct cost.

The cost breakdown given by Watabe and Kondo (1989) for the caisson-based pivoting flap was not fully detailed. Only 74% of the initial capital investment could be associated with a specific cost center; the remainder is presumably distributed among the other categories indicated in Figure 11, and may contain some power conversion equipment as well (only the hydraulic cylinders and motor/generators were broken out in the cited study). It should be further noted that the cost of absorber/reaction structure for this design covers only the concrete caisson. Therefore, the Pendulors are also in the "unknown" balance-of-plant category, but should not account for more than 5% of the total capital cost, based on the amount of steel involved.

**Project Development Risks**

Specific risks associated with the development of a wave power project are associated with uncertainties in capital and operating costs, as well as device performance projections. These risks can be quantified and compared in terms of their impact on the levelized cost of energy.

Uncertainties in equipment costs, energy absorption efficiency, and incident wave power have a direct and obvious impact on the cost of energy. Safety-related risks have a less obvious impact, but can be quantified in the same terms. For example, one of the risks associated with the deployment of offshore wave energy systems is that a device might break free of its mooring and become a drifting navigation hazard. To minimize this risk, the mooring can be substantially over-designed, and chain can be periodically replaced at an interval that is shorter than standard offshore practice. Both of these risk-reduction measures will increase the cost of energy.

To exemplify project development risks for an offshore wave power plant, a cost-of-energy sensitivity analysis was performed on the 30 MWe reference design developed for Makapuu Point (SEASUN Power Systems, 1992).

Uncertainty bands were estimated for eight cost and performance parameters. In order to determine which uncertainties were most significant, each of the parameters was varied from the reference case, one at a time. The results of this sensitivity analysis are presented in Figure 12. The basis for each uncertainty band is described below.

## Cost Uncertainties

**Buoy Fabrication Cost.** The low case is based on a 1986 estimate for ferrocement construction, derived from the scantlings of a prototype tank landing craft (LCT) built and successfully tested by the U.S. Navy in 1945. The high case is based on a 1988 estimate for thin-wall reinforced concrete construction. The difference between the two estimates reflects very different construction methods: use of floating forms and shotcrete laminating in the low case; conventional land-based concrete work in the high case. The reference case was simply chosen as the midpoint between these two estimates (both escalated at an annual inflation rate of 6% to 1990). Of the four cost uncertainties evaluated in this study, buoy fabrication has the biggest impact on the cost of energy, ± 1.3 ¢/kWh.

**Operating and Maintenance Cost.** Marine tanker terminals are the offshore structures that are most similar to floating wave power plants, and for which there is any sort of long-term operational history. Like wave power plants, they are unmanned facilities and must maintain a high availability. The reference case is based on the annual maintenance cost (expressed as a percentage of capital cost) for Exxon's Tembungo Field Single Anchor Leg Mooring, located 93 km offshore Sabba, East Malaysia, in 91 m water depth (Gruy, et al., 1980). Projecting O&M in this manner is not nearly as accurate as a line item estimate, but the latter could not be developed within the resource constraints of the cited study (SEASUN Power Systems, 1992). Therefore, this parameter has a large uncertainty band and a significant impact on the cost of energy, ranging from -1.1 to +1.0 ¢/kWh.

**Indirect Costs.** The cost of land-based general facilities (defined on the previous page) are expected to range from 3% to 15% of the direct cost, while field engineering and home office fees are expected to range from 7% to 15% of the direct cost (Electric Power Research Institute, 1986). The uncertainty band represents the combined additive range of these estimates, with the reference case as the mid-point.

**Deployment Cost.** Mobilization and demobilization charges, as well as lease rates for offshore equipment, are expected to be higher in Hawaii than in the San Francisco Bay area. Relative to the Half Moon Bay Design (SEASUN Power Systems, 1991), the deployment cost for Hawaii was assumed to be the same in the low case, 50% higher in the reference case, and double in the high case.

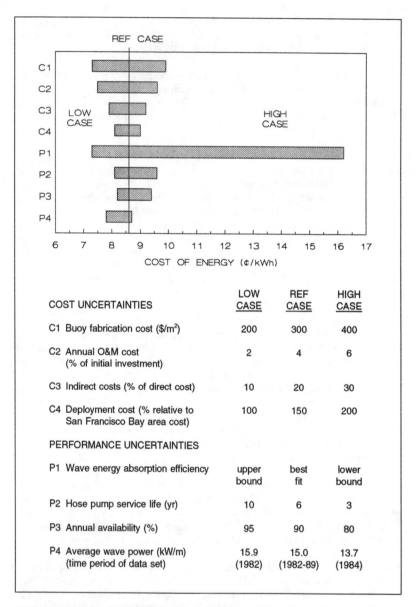

FIGURE 12.  Cost-of-energy sensitivity analysis for 30 MWe reference design of Swedish heaving buoy system off Makapuu Point.  (From SEASUN POWER SYSTEMS, 1992.)

**Performance Uncertainties**

**Wave Energy Absorption Efficiency.** There is a large amount of scatter in the projected absorption efficiency of heaving buoy/hose pump modules in random waves. For a 16-meter diameter buoy in wavelengths typical of the dominant wave period off Makapuu Point the scatter in absorption efficiency data ranges from 25% to 65%. Of all project development risks, this one has the greatest impact on the cost of energy, ranging from -1.3 to +7.6 ¢/kWh.

**Hose Pump Service Life.** The reference case is based on the developers' estimate of 6 years (Svensson, 1985), which was derived from accelerated testing of laboratory-scale prototype hoses. The low case is based on a different set of laboratory tests, conducted by the Harwell Laboratory of the Energy Technology Support Unit in the United Kingdom to evaluate the fatigue characteristics of a polymer tube spring for mooring the submerged Bristol Cylinder wave energy device. Although developed independently of the Swedish hose pump, and for different purposes, the British tube spring is identical to the hose pump in form and function. The Harwell tests indicated a service life in excess of 25 years (Davidson, 1983). Other plant components (connecting hoses subject to bending, upper mooring chain and fittings) are not expected to last this long, however, and would have to be replaced more frequently, incurring the same buoy retrieval and re-deployment costs. Therefore, the low case assumes that the hose pumps, together with these other components, would be replaced at 10-year intervals.

The high case is based on the experience of offshore tanker terminals, where underbuoy hoses have demonstrated a service life of 3-4 years. These hoses are never in tension, but are subject to continual bending. They also carry petroleum products rather than seawater. Nevertheless, they are the only large-diameter submerged hoses for which there is any operational history in the offshore marine environment.

**Annual Maintenance Availability.** The reported availability of turbo-generating machinery in tidal power plants is 96-98% (Carmichael, Adams, and Glucksman, 1986; Rice and Baker, 1987), and that for offshore tanker terminals is 93-96% (Gruy, et al., 1980; Versluis, 1980). Taken together, these give a total availability of 89% to 94% for the underwater habitat and turbine/generator.

Results from the British wave energy program suggested that the annual availability of a wave power plant off the northwest coast of Scotland would be 80% to 90%, based on a variety of devices under consideration at that time. These estimates were derived with a fairly sophisticated computer model that probably represents the best theoretical research in this area (Taylor, 1982).

The reference case approximates the worst reported availability from tidal power/offshore terminal experience, which is similar to the best wave power simulations. The low case approximates the best reported tidal/offshore experience. The high case was based on the worst reported wave power simulations.

**Wave Energy Resource.** The wave data record from offshore Makapuu Point is one of the longest and most complete in Hawaii. Even so, it covers only five complete years and may not represent the long-term wave climate over a 30-year project life. To assess the consequences of this uncertainty, the most and least energetic individual years were used for the low-cost and high-cost cases, respectively.

A similar analysis for the Half Moon Bay design indicated that during the 1981-86 time period, wave energy levels in the north Pacific Ocean were lowest in 1985 and greatest in 1983, an intense "El Nino" year when winter storms devastated much of the California coast (Seymour, 1983). Unfortunately, the wave record from offshore Makapuu Point contains major data gaps during these two extreme years. Of the remaining five years for which seasonally complete records do exist, there is relatively little year-to-year variability. Consequently, the cost-of-energy impact of this parameter is low at Makapuu Point, ranging from -0.8 to +0.1 ¢/kWh. Generally speaking, however, use of only a single year's wave data to forecast device performance entails much greater economic risk than indicated by this particular case.

Several measures can be taken to reduce the uncertainty in cost-of-energy projections, thereby lowering overall project development risk:

- More detailed design work would increase the accuracy of component capital costs. Where a cost uncertainty has an especially large impact on the cost of energy (e.g. buoy fabrication), it may be worthwhile to build a prototype component, which would give a firmer basis for estimating mass-production costs.

- Ocean testing of full-scale prototypes in a wide variety of sea states would reduce the uncertainty in projections of wave energy absorption efficiency, which is the single biggest project development risk.

- Longer-term ocean tests (on the order of years) are required to reduce uncertainties associated with operating and maintenance costs, the required frequency of re-investment in replacement equipment, and annual maintenance availability.

Over-estimation of incident wave power also represents a significant project development risk, and there are several examples of wave energy prototypes that did not perform well due to poor understanding of the local resource. For most projects, only a few years of site-specific measured wave data will be available, and an assessment should be made as to how well those years represent the long-term wave climate. A rough estimate can be obtained by statistical analysis of shipboard observations, which have been archived for well over twenty years in most parts of the world. A more accurate assessment requires a long-term numerical wave hindcast, which has become increasingly affordable in recent years with the development of high-performance desk-top computers and computationally efficient wave hindcasting models (for example, see Earle, 1989).

## Wave Energy Markets and Co-Products

Two applications of wave energy conversion are distributed generation, to meet local demands on the coast, and central stations, designed to export bulk power inland to the main utility grid. Distributed generation plants are likely to be deployed before central stations, and their commercial appeal can be enhanced by a variety of co-products. In the long term, wave power may find application in current proposals for the offshore utilization of sea space. Various market development scenarios are outlined in the last part of this section.

### Distributed Generation

Distributed generation is defined as a wave power plant designed to meet the electric power demands of a coastal load center. An alternating-current (AC) submarine power cable would be landed into an existing coastal substation, and the necessary transformers and switchgear would be provided to connect to the local distribution system. Export of power inland would be incidental only to variations in local demand, and it is assumed that the installed wave generating capacity would be sufficiently small so as not to require any reinforcement of existing onshore transmission lines.

Coastal towns in high latitudes seldom require summer air-conditioning, and where natural gas service is not available for winter heating, the seasonal variation in electricity demand is well-matched to that of the wave energy resource. The remoteness of the coast also makes it attractive for "mini-utilities", whereby winter peak load growth is met by building small, local wave power plants rather than additional transmission lines from the inland grid. A survey of coastal substations in northern California suggests that the useful size of such a wave power plant would range from one to ten megawatts (SEASUN Power Systems, 1991). It should be noted that these comments apply to large mainland utility grids; for many island applications, plant sizes on the order of tens to hundreds of kilowatts would be more appropriate.

In addition to meeting coastal load growth, a few megawatts of distributed wave generation capacity may benefit the regional transmission network by providing grid stability and voltage support, particularly in remote coastal areas with load centers at the end of a long transmission circuit. A recent study by Pacific Gas and Electric Company has suggested that small-scale photovoltaic (PV) plants would provide avoided-cost benefits to the local distribution system comparable to the value of the PV plant itself (Shugar, 1990). There is every reason to believe that a small wave power plant could provide the same benefits.

As important as distributed wave generation may be at the local or regional level, its impact on an entire grid may be insignificant. For industrialized countries, any significant development of the wave energy resource probably will require the deployment of central generating stations.

**Central Station Generation**

A major consideration for central station applications is inland transmission of the power once it lands ashore. A second consideration is managing the power so it can be most effectively combined with other generating plant.

In many regions of the world, good wave energy resources are located along coastal stretches that are relatively unpopulated and remote from main-grid load centers (examples include northern California and the northwest coast of Scotland). Even where a utility grid extends to such a coastal area, it is usually not designed to handle large amounts of power. Therefore inland export of wave power from coastal substations via existing transmission lines is likely to be severely limited.

Another possible inland route for wave power is surplus transmission capacity at thermal power plants located on the coast for access to cooling water; however, these plants may be far removed from the best wave energy resources. Furthermore, much of the surplus may be unavailable, due to other utility requirements for these lines.

Therefore, substantial development of the wave energy resource for bulk power generation will require reinforcement or extension of the onshore utility grid. Recent studies in the United Kingdom suggest that for a 2,000 MWe wave power station off the Outer Hebrides Islands, reinforcement of the onshore grid will be as expensive, if not more, than sea-to-shore power transmission (Thorpe, 1992)

Once the power reaches the central grid, there are further difficulties. Although waves are more persistent than the winds that generate them, output from a wave power plant can fall below rated capacity at irregular intervals lasting from a few hours to a number of days. Assume for a moment that the cost of energy from a central wave power station was comparable to that from thermal plant, and that the latter was used to make up for hourly or daily shortfalls in wave power. It is not practical to start up and shut down large coal-fired plants over this time scale. Either smaller gas- or oil-fired plants (which have higher energy costs than coal) must be available, or a large coal-fired plant would have to be kept on part load (spinning reserve), which is relatively inefficient. Alternatively, one could integrate long-term energy storage, such as an onshore pumped hydro station, with the wave power plant.

Modern wave forecasting models exist that can predict wave conditions 24 to 48 hours in advance with reasonable accuracy, easing the dispatch problem. This does not eliminate the need, however, for back-up generation or long-term energy storage.

Siting two or three smaller wave power plants at widely spaced intervals will provide a greater level of firm capacity than siting the same amount of wave generating plant in one location. Another way to increase wave power's baseload capacity contribution is to deploy more absorbers per unit generator rating. While this would be economical if the absorbers were a small part of total plant cost, this is not the case with the present state of the art, as shown in Figure 12.

A possibly more effective use of central station wave power would be the production of hydrogen as a winter heating fuel. This provides a good match to seasonal demand, eliminates the dispatchability problem, and avoids the environmental and economic costs of reinforcing or extending the onshore utility grid. Furthermore, a recent study by Pacific Gas and Electric Company suggests that hydrogen for heating is a more economical carrier of renewable energy than either hydrogen in fuel cells or electric power transmission with pumped storage (Braun, Suchard, and Martin, 1990). Only where no storage is required, as with naturally dispatchable photovoltaic plant, does electric power transmission become the more economical alternative. Wave energy processes must be proven in distributed generation applications, however, before central station applications (electricity or hydrogen) can become a reality.

## Wave Energy Co-Products

A variety of products can be generated by wave energy systems, incidental to the generation of electricity, including:

- Seawater renewal for closed-pond aquaculture
- Fresh water for household use, livestock watering, and crop irrigation
- Breakwater protection for ports and harbors

Co-product potential varies among different wave energy technologies, depending on whether they are fixed or floating devices, and whether their working fluid is air, hydraulic oil, or seawater. Some devices could be modified to generate electricity and all three co-products in a single plant. For example, the closed-circuit hydraulic system of a caisson-based pivoting flap could be replaced with a high-pressure seawater system. Some of the pumped seawater could be diverted to onshore aquaculture ponds, while the rest would enter a reverse osmosis (RO) module. At a seawater pressure of 800 psi (5.5 MPa), only 20% of this flow would pass through the RO membrane as fresh water; the remainder could be discharged through a Pelton turbine to generate electricity. Such a system, however, is now a matter of speculation. The following text describes co-product applications that have been studied for state-of-the-art wave energy devices.

**Aquaculture.** The reservoir of a Tapered Channel power plant does not provide long-term storage, but smooths the input from one high-energy wave group to the next. The reservoir at Toftestallen has an area of 5,500 m$^2$, while the 350 kWe turbine/generator has an operating head of 3 m and a flow rate of 14 to 16 m/sec (Norwegian Royal Ministry of Petroleum and Energy, 1987). Assuming a turbine/generator efficiency of 90%, the upper meter of the reservoir is replaced once every six minutes, which represents a tremendous rate of seawater renewal. The Crown Agents of London have estimated that the value of fish farmed in a Tapered Channel reservoir "may cover all capital costs of wave collectors, converters and basin" (Mehlum, et al., 1990).

Deep oceanic water is cold, nutrient-rich, and relatively pathogen-free, which makes it an ideal resource for aquaculture, providing ease of temperature regulation, enhanced productivity, and reduced exposure to infection. Its use for open-ocean aquaculture has been proposed in connection with ocean thermal energy conversion (OTEC), but wave-powered pumps also have been proposed for the same purpose (Liu and Chen, 1991).

**Fresh Water.** Although the vast majority of wave energy systems are designed for electric power generation, a heaving buoy device (DELBUOY) was developed by researchers at the University of Delaware for the desalination of seawater by reverse osmosis. Since 1982, an evolving series of full-scale prototypes has been deployed off the southwest coast of Puerto Rico, producing fresh water at a continuous rate of 950 liters (250 gallons) per day from a single buoy (Hicks, Pleass, and Mitcheson, 1988).

DELBUOY consists of a 2.1 m diameter buoy tethered to a seafloor anchor by a single-acting hydraulic cylinder, which has a bore diameter of 4 cm. This gives a pressure amplification ratio of approximately 2750:1, such that relatively small waves can generate pump pressures as high as 5.5 MPa (800 psi). Six buoy/pump moorings supply one RO module that delivers 5,700 liters (1,500 gallons) of fresh water per day, in waves 1 m high, having a period of 3-6 seconds. Such waves are typical in the Caribbean Sea and along the arid coastal regions of Africa, where average incident wave power is on the order of 1-10 kW per meter of shoreline.

**Breakwater Protection.** Since caissons reflect all wave energy that they don't absorb, there is the possibility of a distinct low-energy "shadow" developing behind the plant, depending on how closely the individual caissons are spaced. This maximizes their potential environmental impact on longshore sediment transport. It is this reflective feature of caissons, however, that make them ideal candidates for combining wave energy generation with breakwater protection at coastal harbors. In such cases their visual appearance would probably not be considered intrusive, since local scenery has already been altered by the existing harbor development.

Even in remote coastal areas, an occasional wave energy breakwater might be acceptable for creating a harbor of refuge for small craft in the event of sudden storms, medical emergencies, or engine problems. The co-production of fresh water and electricity would have obvious advantages in such an application.

It should be noted, however, that the potential market for wave energy breakwaters is relatively small. For example, Pacific Gas and Electric Company (PG&E) has an open ocean coast approximately 1,000 km in length. Yet a recent survey identified only two existing harbors that needed additional breakwater protection, and five coves that could be developed into harbors of refuge. The potential wave generating capacity at these seven sites was estimated to be only 12.5 MWe (SEASUN Power Systems, 1991), which is insignificant in terms of PG&E's grid-wide demand. Nevertheless, if even the smallest of these (700 kWe) was developed, it would more than double the world's present wave generating capacity. Therefore, breakwaters are an important stepping stone in the commercialization of wave power.

## Offshore Utilization of Ocean Space

Very large floating structures (VLFS) have been proposed for a variety of offshore applications, including airports, space launch facilities, naval bases, living and recreation complexes, and production of energy-intensive materials and fuels. Barge-like VLFS modules are less costly than semisubmersibles, but require protection from wave action (Chow, et al., 1991). Fabricating the perimeter of such a VLFS to accommodate wave energy absorbers such as oscillating water columns, pivoting flaps, or flexible bags should not add significantly to its total construction cost.

The large inertia of the VLFS provides the necessary reaction point for these absorbers, which in turn act to remove energy from the wave before it passes under the platform. It is important to note that much of this energy is ultimately absorbed by the VLFS via structural foundations for power conversion equipment, but these may be less costly than designing the platform to withstand large sagging and hogging loads associated with undampened waves.

As previously noted, reaction structures are the single largest capital cost item for caisson-based wave energy devices and the SEA Clam. If the fabrication cost of the VLFS is supported by revenue derived from its main function, then wave energy costs for power conversion equipment and other balance-of-plant may be half or less than that now projected for systems connected to onshore utility grids (see Figure 12).

The development of VLFS-based offshore wave power plants could lead to the large-scale production of hydrogen for mainland electric power generation and as a heating or transportation fuel. If so, wave energy has the ultimate potential to supply a large portion of global energy demand, as explained below.

A steady wind blowing over water imparts its energy to waves, which grow in size until they become unstable and break. Shorter-period waves reach limiting steepness first, while longer-period waves continue to grow. Longer-period waves also have a higher group velocity (the speed at which wave energy propagates across the ocean surface), and for a given wind speed, waves longer than a certain period travel faster than the wind. Once waves of this period have reached limiting steepness, the sea can absorb no more wind energy and is said to be fully developed. At this point, energy input from the wind is offset by energy losses due to wave breaking.

If energy is absorbed from a fully developed sea by a wave power plant, however, the waves can absorb more energy from the wind than otherwise. Given sufficient fetch downwind of the plant, the waves will again become fully developed, regaining the energy that was absorbed. To use an agricultural analogy, waves re-grow as they are "grazed", and the overall productivity of the sea surface (in terms of rate of energy yield, or power) increases. If such cropping can be achieved economically on a worldwide scale, then the global wave energy resource base exceeds that of ocean thermal energy conversion (OTEC) and is comparable to civilization's present power demand (Isaacs and Schmitt, 1980).

The persistent trade winds of the tropics provide a wave climate in which such cropping is a realistic proposition. Mathematical modeling indicates that at a wind speed of 10 m/sec, wave regrowth between power plants spaced 50 km apart along a 600 km fetch increases wave energy yield by 40% (Claeson and Sjostrom, 1991).

The ultimate prospect of large-scale production of wave-generated hydrogen suggests certain criteria for wave energy conversion devices that might be targeted for accelerated research and development during the next five years. One criterion is an absorber configuration that can be easily integrated into VLFS perimeter construction. Another is the use of high-pressure seawater as a working fluid. Efficient electrolyzers require fresh water, and DELBUOY has demonstrated that with a sufficiently large ratio of absorber area to pump cross-section, seawater can be desalinated directly via reverse osmosis, without an electrical intermediary step. Thus, high-pressure seawater systems have both long-term commercial potential for efficient hydrogen production offshore, and near-term commercial potential for cogeneration of fresh water and electricity at remote island and coastal locations.

## Market Development

The near-term market for wave power is clearly in the Pacific Ocean, which has an energetic wave climate (despite its name) and a large number of island nations where the need for small-scale electric power and fresh water is great. If electrified at all, these islands are typically served by old diesel units, some dating back to the second world war, operating at the far end of the oil distribution network.

Where coastal land is not at a premium, and where a reservoir can be built without extensive blasting or dam construction, Norwave's Tapered Channel is the most economical wave energy system developed to date, and is expected to capture a fair share of the Pacific market. Japanese caisson-based systems are also expected to enter this market within the next five years.

The commercial export of offshore systems is expected to lag behind that of land- and caisson-based systems. Significant improvement in cost and performance will not occur until a floating device is successfully demonstrated at full scale for a period of years. Commercial sales will not occur without a successful demonstration, and without sales, the developers of such systems will not have the financial resources or operating experience necessary to make significant improvements.

Based on the foregoing text, it would seem that for industrialized nations, commercial development of wave power beyond a few tens or hundreds of megawatts depends on the construction and operation of a demonstration plant for performance and endurance testing of offshore components. To the extent that reinforcement or extension of onshore utility grids is prohibitively expensive or environmentally unacceptable, large-scale wave power development may also depend on the existence of an onshore infrastructure for the handling, storage and distribution of hydrogen.

Without the long-term component testing afforded by a demonstration plant, it is doubtful whether any sort of offshore wave energy system would become economically feasible in the foreseeable future. Under this scenario, wave power development in industrial countries will be largely limited to caisson-based systems deployed as breakwaters at locations where their environmental impact is consistent with the existing level of onshore harbor development. This would probably amount to only tens of megawatts worldwide.

Without the development of an onshore hydrogen infrastructure, utilization of wave power will be limited by the economic and environmental costs of reinforcing or extending the utility grid and wave energy's lack of dispatchability. Under this scenario, it is expected that wave energy would be used primarily for distributed generation, as described previously. This would probably amount to only hundreds of megawatts worldwide.

With both drivers in place, however, it is conceivable that industrial countries could meet a large part of their energy needs from wave power. Under this scenario, tens of megawatts of caisson-based capacity would be deployed over the next decade. These would be demonstration plants designed primarily for harbor protection, but which would incorporate components used in various offshore wave energy systems, for purposes of long-term endurance testing.

By the end of this century, sufficient component sea time would be accumulated to enable confident projections of the cost and performance of offshore wave power plants. These would then be commercially deployed: first in island applications where the competing technology is diesel generation; later, in distributed "mini-utilities" along mainland coasts, where the competing technology is extension of the central grid.

Meanwhile, an onshore hydrogen infrastructure would have been developed, fueled initially by surplus hydroelectric generation. The incentive for this development might come from depletion of fossil methane reserves (the feedstock for existing hydrogen production by steam reformation), coupled with legislation designed to reduce carbon dioxide emissions.

This does not require a full-blown "hydrogen energy economy", where hydrogen is also used for transportation and electric power generation. It does require, however, sufficient infrastructure that hydrogen can be used as a replacement fuel for major heating applications now served by natural gas.

Should an onshore infrastructure develop to receive hydrogen, wave energy has several advantages over competing technologies to provide it. Unlike steam reformation of methane or coal gasification, hydrogen production via electrolysis produces no carbon dioxide. Water supply would not be a problem for wave-powered electrolysis but could be a limiting factor in those inland areas having good solar or wind resources. While these renewables could gain access to water by moving them offshore, wave energy makes much more efficient use of sea space (SEASUN Power Systems, 1991).

## Cost-of-Energy Comparison with Other Energy Options

In the above market development scenario, wave power must be commercially successful in distributed generation before it can be seriously considered for central station use. The likelihood of this development depends a great deal on the cost of wave-generated electricity relative to that generated by other technologies.

This section examines wave energy's present state of economic competitiveness. As before, EPRI financial parameters were used to compute levelized energy costs in constant 1990 U.S. dollars (see Appendix for details).

Looking first at the export market, Table 4 compares the projected cost of energy for various projects in the South Pacific Ocean. The Stewart Island examples are from a study of central electricity supply options for the island's 500 inhabitants (Royds Garden Limited and Worley Consultants, 1985). Although New Zealand and eastern Australia are not considered viable candidate sites for land-based OTEC plants, many nearby island nations are, including Fiji, Tonga, and Samoa (Vega, 1992).

The projected cost of wave energy in such small-island applications is 4 to 5 ¢/kWh lower than hydro or diesel, and less than half the cost of a land-based OTEC plant. It should be noted that King Island is located south of the Australian mainland, where it is fully exposed to high-energy Southern Ocean swell. As this swell travels north, it becomes more widely dispersed and its energy density decreases. In the same direction, however, trade wind wave energy levels increase. Overall, incident wave power in the region framed by Vanuatu, Tongatapu, Rarotonga, and Tuvalu is expected to average 15-20 kW/m annually (Barstow, Soras, and Selanger, 1991). As indicated by Norwave's Java project (Table 1) and the heaving buoy design for Makapuu Point (Table 3), this is still an adequate resource for competitive wave energy costs.

While the comparisons of Table 4 are applicable to smaller projects, on the order of a few megawatts, there are many Pacific islands with a well-established utility grid and large demand growth, such that projects on the order of tens to hundreds of megawatts are warranted. Representing this market, Table 5 presents cost-of-energy comparisons for the densely populated island of Oahu, Hawaii. Although wave power is still competitive with OTEC there, it is more costly than coal by 2 to 3 ¢/kWh.

It should be noted that the coal-fired plant at Barbers Point was commissioned early in 1992, so that cost and performance projections for this project are quite reliable. Less certain are the wave power plant data. Although based on vendor estimates and existing technology, the original design was for deployment off northern California, later adapted to Hawaii, as described previously. Even less certain are the OTEC data, which include cost reductions associated with engineering developments beyond the present state of the art, particularly in cold water pipe design (Vega, 1992). There is also a wide range in sub-system cost estimates among different OTEC design studies (Zangrando, 1992), which adds further uncertainty to cost projections based on engineering development of any particular sub-system.

TABLE 4. Wave, OTEC, small hydro, and diesel comparison in the southwest Pacific Ocean. [a]

| Option: | WAVE | OTEC | HYDRO | DIESEL |
|---|---|---|---|---|
| Technology: | Tapered Channel | open-cycle, w/ 2nd stage condenser [b] | 10 m dam height, 80 m dam length, 8 km² reservoir | water-cooled w/ heat rate of 11,400 BTU/kWh |
| Location: | King Island (Australia) | unspecified | Stewart Island (New Zealand) | Stewart Island (New Zealand) |
| Energy resource: | Southern Ocean swell at 30-32 kW/m | surface water at 26°C, deep water at 4°C | mean river flow of 4.2 m³/sec, hydraulic head of 34.5 m | bulk diesel fuel w/ 37,000 BTU/liter heat content |
| Plant capacity: | 1.5 MWe | 1.256 MWe (net) | 1.2 MWe (2 x 600 kWe) | 0.9 MWe (3 x 300 kWe) |
| Annual output: | 9 GWh | 8.8 GWh | 5 GWh [c] | 5 GWh [c] |
| Capacity factor: | 68% | 80% | 48% | 64% |
| Initial investment: | $3,290-3,840/kWe | $23,000/kWe | $4,290/kWe [d] | $600/kWe |
| Re-investment: | none | none | none | $268/kWe at 10 yr |
| Annual O&M cost: (non-fuel) | $33/kWe | $391/kWe | $25/kWe | $71/kWe incl. variable |
| Annual fuel cost: | none | none | none | 9.7 ¢/kWh [e] |
| Plant service life: | 40 yr | 30 yr | 40 yr | 20 yr |
| Cost of energy: | 6.2-7.2 ¢/kWh | 14.9 ¢/kWh [b] | 11.3 ¢/kWh | 12.6 ¢/kWh |

[a] A construction time of one year was used for all projects. Costs are in 1990 constant dollars. See Appendix for financial parameters and currency exchange rates. (Data from MEHLUM, 1990; VEGA, 1992; ROYDS GARDEN LIMITED AND WORLEY CONSULTANTS, 1985.)

[b] Co-production of fresh water at an average rate of 4,000 m³/day. Cost of energy includes credit of 26.5 ¢/kWh, based on an assumed water selling price of $1.6/m³ ($6/kgal).

[c] Forecast demand ten years after hydro or diesel plant comes on line.

[d] Does not include cost of transmission (21 km overhead, 7 km submarine), access road, or 0.6 MWe (2 x 300 kWe) standby diesel plant in Oban.

[e] Equivalent to a bulk diesel fuel price of 31.4 ¢/liter ($50 per 42-gal barrel).

TABLE 5.  Wave, OTEC, and coal comparison on Oahu, Hawaii. [a]

| Option: | WAVE | OTEC | COAL |
|---|---|---|---|
| Technology: | Swedish heaving buoy | Land-based, hybrid cycle [b] | Atmospheric fluidized bed w/ heat rate of 10,300-10,800 BTU/kWh |
| Location: | 10 km offshore Makapuu Point, 80 m depth | unspecified | Barbers Point |
| Energy resource: | north Pacific swell and trade wind waves at 15 kW/m | warm water at 26°C, deep water at 4°C | Indonesian coal |
| Plant capacity: | 30 MWe | 40 MWe (net) | 180 MWe |
| Annual output: | 111 GWh | 280 GWh | 1,420 GWh |
| Capacity factor: | 42% | 80% | 90% |
| Initial investment: | $1,970/kWe | $8,700/kWe | $2,130/kWe |
| Re-investment: | $173/kWe every 6 yr | none | unknown |
| Annual O&M cost: (non-fuel) | $79/kWe | $131/kWe | 0.8-1.2 ¢/kWh incl. fixed |
| Annual fuel cost: | none | none | 1.8-2.4 ¢/kWh [c] |
| Plant service life: | 30 yr | 30 yr | 40 yr |
| Construction time: | 1 yr | 2 yr | unspecified [d] |
| Cost of energy: | 8.6 ¢/kWh | 9.3 ¢/kWh [b] | 5.4-6.4 ¢/kWh |

[a]  Costs are in 1990 constant dollars.  See Appendix for financial parameters.  (Data from SEASUN POWER SYSTEMS, 1992; VEGA, 1992; HEMPHILL, 1992.)

[b]  Closed-cycle ammonia loop for electric power generation; open-cycle 2nd stage with surface condenser for production of fresh water at an average rate of 62,000 m³/day. Cost of energy includes credit of 6.5 ¢/kWh, based on an assumed water selling price of $0.8/m³ ($3/kgal).

[c]  Based on a projected coal cost of $1.70-$2.20/million BTU.

[d]  Initial investment includes interest during construction and other pre-commissioning costs.

Moving from island to mainland markets (Table 6), it is clear that levelized energy costs must drop by a factor of two before wave power will be competitive with wind power or coal-fired generation. As explained previously, capacity factor is lower for the Swedish heaving buoy design in northern California than in Hawaii. Even though wave energy levels are greater off Half Moon Bay, the energy is contained in longer-period waves and is not absorbed as efficiently as the shorter-period trade wind wave energy off Makapuu Point.

TABLE 6. Wave, wind, and coal comparison in northern California. [a]

| Option: | WAVE | WIND | COAL |
|---|---|---|---|
| Technology: | Swedish heaving buoy | two-hundred 200 kWe turbines | pulverized coal w/ heat rate of 10,000 BTU/kWh |
| Location: | 24 km offshore Half Moon Bay | Altamont Pass | Montezuma |
| Energy resource: | north Pacific swell at 25 kW/m | unspecified | unspecified |
| Plant capacity: | 30 MWe | 40 MWe | 460 MWe |
| Annual output: | 76.3 GWh | 87.6 GWh | 2,820 GWh |
| Capacity factor: | 29% | 25% | 70% |
| Initial investment: | $2,100/kWe | $910/kWe | $2,500/kWe |
| Re-investment: | $143/kWe every 6 yr | none | none |
| Annual O&M cost: (non-fuel) | $84/kWe | $1.2/kWe and 1.2 ¢/kWh | $55/kWe and 0.5 ¢/kWh |
| Annual fuel cost: | none | none | 1.9 ¢/kWh [b] |
| Plant service life: | 30 yr | 25 yr | 30 yr |
| Construction time: | 1 yr | 1 yr | 5 yr |
| Cost of energy: | 13 ¢/kWh | 6.0 ¢/kWh | 7.9 ¢/kWh |

[a] Costs are in 1990 constant dollars. See Appendix for financial parameters. (Data from SEASUN POWER SYSTEMS, 1991; DOYLE, EYER, AND HAY, 1988.)

[b] Based on a projected coal cost of $1.94/million BTU.

OCEAN ENERGY RECOVERY

None of the regions examined thus far are subject to large tides. Comparison of wave and tidal power is possible in the United Kingdom, however, where good resources exist for both technologies. As shown in Table 7, an important factor in this comparison is size. While the 12.5 MWe circular SEA Clam is competitive with tidal power on the order of a few tens of megawatts or less, it is clearly not competitive with a several-thousand megawatt plant such as that proposed for the Severn Estuary.

TABLE 7. Wave and tidal power comparison in the United Kingdom. [a]

| Option: | WAVE | TIDE | TIDE | TIDE |
|---|---|---|---|---|
| Technology: | Circular SEA Clam | Salford Transverse Oscillator | single-effect, ebb generation | single-effect, ebb generation, w/ flood pumping |
| Location: | 5 km offshore Lewis | Callanish, Lewis | Langstone Harbor | Severn Estuary |
| Energy resource: | north Atlantic swell at 52 kW/m | 2.9 m tidal range | 3.1 m tidal range | 11 m tidal range |
| Barrage length: | not applicable | 340 m | 550 m | 15.9 km |
| Reservoir area: | not applicable | 2.7 km² | 19 km² | 480 km² |
| Plant capacity: | 12.5 MWe | 3.13 MWe | 24.3 MWe | 8,640 MWe |
| Annual output: | 27.9 GWh | 4.5 GWh | 53.2 GWh | 17,000 GWh |
| Capacity factor: | 25% | 16% | 25% | 22% |
| Initial investment: | $2,210/kWe | $2,060/kWe | $3,480/kWe | $1,640/kWe |
| Re-investment: | none | none | $1,200/kWe every 40 yr | $178/kWe every 30 yr |
| Annual O&M cost: | $88/kWe | $43/kWe | $41/kWe | $14/kWe |
| Plant service life: | 25 yr | 25 yr | 120 yr | 120 yr |
| Construction time: | 1 yr | 1 yr | 5 yr | 8 yr [b] |
| Cost of energy: | 15 ¢/kWh | 19 ¢/kWh | 21 ¢/kWh | 11 ¢/kWh |

[a] Costs are in 1990 constant dollars. See Appendix for financial parameters and currency exchange rates. (Data from BELLAMY, 1992; CARNIE, 1989; YARD LTD, 1989; DEPARTMENT OF ENERGY, CENTRAL ELECTRICITY GENERATING BOARD, AND SEVERN TIDAL POWER GROUP, 1989.)

[b] Seven years to barrage closure and first commercial power generation, with a further two years to completion and full output.

For electric power generation, then, it appears that wave power is highly competitive with other renewables and with fossil fuels in island markets where capacity additions on the order of a few megawatts are contemplated. On the industrialized mainland and large islands where high demand growth warrants capacity additions of a few hundred megawatts, wave power is not competitive with coal. Among ocean energy options, however, only the largest tidal power stations have lower energy costs.

Regarding fresh water production, there is little economic data available for wave power. While DELBUOY is close to being a commercial product, its output is relatively low, approximately 5.7 m³/day (1,500 gal/day) for a six-buoy system in a trade wind wave climate. As shown in Table 8, such a system is not well matched to the levels of demand on developed islands. In developing countries, however, even a few cubic meters of fresh water per day would significantly improve basic living conditions in communities now without a reliable water supply.

TABLE 8.  Electricity and fresh water demand on four Hawaiian islands. [a]

| Island: | Molokai | Kauai | Maui | Oahu |
|---|---|---|---|---|
| Resident population: | 6,700 | 47,600 | 81,100 | 830,600 |
| Total annual electricity demand: | 25.4 GWh | 270 GWh | 620 GWh | 5,804 GWh |
| Average per capita electricity demand <br> • all uses: <br> • residential: | 10.4 kWh/day <br> 4.3 | 15.5 kWh/day <br> 5.6 | 20.9 kWh/day <br> 6.8 | 19.1 kWh/day <br> 5.0 |
| Total annual fresh water demand [b]: | 17.6x10⁶ m³ | 314x10⁶ m³ | 646x10⁶ m³ | 555x10⁶ m³ |
| Average per capita fresh water demand <br> • all uses [b]: <br> • domestic: <br> • agricultural: | 7.2 m³/day <br> 0.8 <br> 6.4 | 18.1 m³/day <br> 0.9 <br> 16.5 | 15.8 m³/day <br> 0.7 <br> 15.0 | 1.8 m³/day <br> 0.6 <br> 0.9 |
| Fresh water demand per unit electricity demand (all uses): | 0.7 m³/kWh | 1.2 m³/kWh | 0.8 m³/kWh | 0.1 m³/kWh |

[a]  Population estimate is as of 1 July 1987, electricity demand is for 1987, fresh water demand is for 1985. (Data from STATE OF HAWAII, 1988).

[b]  Does not include water diverted for hydroelectric power generation (which is significant on Kauai), since this does not represent a potential market for desalinated seawater.

The development of large wave-powered desalination systems has been little explored, yet the potential energy benefits of such a technology are great. Conventional reverse osmosis (RO) plants consume 5.8 kWh per m³ of product (22 kWh/kgal), even when fitted with energy recovery systems (Block, 1991). Using conventional RO to supply only 10% of the fresh water demand on Molokai would increase total electricity consumption on that island by 40%; the impact would be even greater on Maui (60%) and Kauai (67%). In arid coastal regions and small islands, where fresh water supplies are inherently limited, a growing population's need for water could well create a worse energy crisis than its demand for electric power.

Assuming that competitive prices for fresh water and electricity are $.80/m³ ($3/kgal) and $.08/kWh, respectively, then the market for fresh water is at least comparable in dollar value to that for electricity on an urbanized island such as Oahu, and many times greater on islands with extensive ranching and farming. It is difficult to escape the conclusion that wave power may be commercialized sooner if fresh water, rather than electricity, is viewed as the primary product. Such a development, if realized, will owe much to the pioneering work of Hicks, Pleass, and Mitcheson (1988).

## References

B. Hojlund Rasmussen, 1988.  *Projektering af et bolgedraftvaerk, delfase B, modelforsog og skitseprojekt.* Copenhagen, Denmark:  B. Hojlund Rasmussen, June 1988 (in Danish).

Barstow, S.F., P.-E. Soras, and K. Selanger, 1991.  Wave energy resource mapping for Asia and the Pacific.  In *Oceans'91 Proceedings*, pp. 385-390 (Vol. 1).  New York, New York:  Institute of Electrical and Electronics Engineers.

Bellamy, Norman, 1992.  Telephone conversation with author, 10 July.

Block, D.L., 1991.  *Cost of Desalinated Water from Reverse Osmosis and Photovoltaic-Powered Reverse Osmosis.*  Cape Canaveral, Florida:  Florida Solar Energy Center, FSEC-RR-20-91.

Bonke, K., and N. Ambli, 1987.  Prototype wave power stations in Norway.  In *Utilization of Ocean Waves - Wave to Energy Conversion*, edited by M. E. McCormick and Y. C. Kim, pp. 34-44.  New York, New York:  American Society of Civil Engineers.

Braun, G. W., A. Suchard, and J. Martin, 1990.  Hydrogen and electricity as carriers of solar and wind energy for the 1990s and beyond.  In *Proceedings of the IEA Fifth Symposium on Solar High-Temperature Technologies*, (in press).

Carmichael, A.D., E.E. Adams, and M.A. Glucksman, 1986. *Ocean Energy Technologies: The State of the Art.* Palo Alto, California: Electric Power Research Institute, EPRI AP-4921.

Carnie, Colin, 1989. Letter to author, 10 November.

Chino, H., K. Nishihara, and Y. Nakakuki, 1989. *Air Conditioning With Heat Pump.* Honolulu, Hawaii: Department of Business and Economic Development, Energy Division (reprint on file).

Chow, P.Y., T.Y. Lin, H.R. Riggs, and P.K. Takahashi, 1991. Engineering concepts for design and construction of very large floating structures. In *Proceedings, First International Workshop on Very Large Floating Structures,* edited by R.C. Ertekin and H.R. Riggs, pp. 97-106. Honolulu, Hawaii: University of Hawaii at Manoa.

Claeson, L., and B.-O. Sjöstrom, 1991. Optimal wave power plant spacing in the tropical north Pacific Ocean. In *Oceans'91 Proceedings,* pp. 547-549 (Vol. 1). New York, New York: Institute of Electrical and Electronics Engineers.

Davidson, R., 1983. *The Mechanical Characteristics of Spirally Reinforced Moorings and Tube Pumps.* Oxfordshire, United Kingdom: Energy Technology Support Unit, Harwell Laboratory, WV-1551-P55b.

Department of Energy, Central Electricity Generating Board, Severn Tidal Power Group, 1989. *The Severn Barrage Project: General Report.* London, United Kingdom: Department of Energy, Her Majesty's Stationery Office, Energy Paper Number 57.

Doyle, J. F., J. M. Eyer, and G. A. Hay, 1988. *Scenario Evaluation and Research Choice (SEARCH) Study - Phase I.* San Ramon, California: Pacific Gas and Electric Company, Department of Research and Development, 005-88.1.

Earle, M. D., 1989. Microcomputer numerical ocean surface wave model. *Journal of Atmospheric and Oceanic Technology,* Vol. 6, pp. 151-168.

Electric Power Research Institute, 1986. *Technical Assessment Guide, Volume 1: Electricity Supply-1986.* Palo Alto, California: Electric Power Research Institute, P-4463-SR, Vol. 1.

Electric Power Research Institute, 1987. *Technical Assessment Guide, Volume 3: Fundamentals and Methods, Supply-1986.* Palo Alto, California: Electric Power Research Institute, P-4463-SR.

Gruy, R.H., W.L. Kiely, K.I. Pederson, W.R. Wolfram, and R.D. Swann, 1980. Five years experience with the first deepwater SALM. In *Proceedings of the 12th Annual Offshore Technology Conference,* Paper No. 3804.

Hemphill, Robert, 1992. Telephone conversation with author, 24 January.

Hicks, D. C., C. M. Pleass, and G. R. Mitcheson, 1988. DELBOUY: Wave-powered desalination system. In *Oceans'88 Proceedings*, pp. 1049-1055 (Vol. 3). New York, New York: Institute of Electrical and Electronics Engineers.

Isaacs, J. D., and W, R. Schmitt, 1980. Ocean energy: forms and prospects. *Science*, Vol. 207, pp. 265-273.

Ishii, S., T. Miyazaki, Y. Masuda, and G. Kai, 1982. Reports and future plans for the Kaimei project. In *Proceedings of the Second International Symposium on Wave Energy Utilization*, edited by H. Berge, pp. 305-321. Trondheim, Norway: Tapir Publishers.

Jeffreys, E., 1985. New power plants harness the Atlantic waves. *Veritas*, March/May, pp. 28-30.

Kondo, H., and T. Watabe, 1990. Optimal depth for nearshore wave power extractors of fixed type. In *Proceedings of the International Conference on Ocean Energy Recovery*, edited by Hans Jurgen-Krock, pp. 59-66. New York, New York: American Society of Civil Engineers.

Liu, C.C.K., and H.H. Chen, 1991. Conceptual design and analysis of a wave-driven artificial upwelling device. In *Oceans'91 Proceedings*, pp. 406-412 (Vol. 1). New York, New York: Institute of Electrical and Electronics Engineers.

Mehlum, Even, 1990. Telephone conversation with author, 9 November.

Mehlum, E., 1991. Commercial Tapered Channel wave power plants in Australia and Indonesia. In *Oceans'91 Proceedings*, pp. 535-538 (Vol. 1). New York, New York: Institute of Electrical and Electronics Engineers.

Mehlum, E., P. Anderssen, T. Hysing, J.J. Stamnes, O. Eriksen and F. Serck-Hanssen, 1990. The status of wave energy projects and plants in Norway. Part 2: Norwave TAPCHAN - a commercial overview.. In *Proceedings: Oceans'89 Special International Symposium - A Global Review of the Development of Wave Energy Technologies*, edited by D. W. Behrens and M. A. Champ, pp. 29-36. San Ramon, California: Pacific Gas and Electric Company, Department of Research and Development, 007.6-90.8.

Nielsen, K., and C. Scholten, 1990. Planning a full-scale wave power conversion test: 1988-1989. In *Proceedings of the International Conference on Ocean Energy Recovery*, edited by Hans-Jurgen Krock, pp. 111-120. New York, New York: American Society of Civil Engineers.

Norwegian Royal Ministry of Petroleum and Energy, 1987. *Norwegian Wave Power Plants 1987.* Oslo, Norway: Royal Ministry of Petroleum and Energy.

Reuter News Service, 1989. Norwegian wave energy plant blown out to sea in storm. *International Solar Energy Intelligence Report,* 27 January, p. 20.

Rice, R.G., and G.C. Baker, 1987. The Annapolis experience. In *Oceans'87 Proceedings,* pp. 391-396 (Vol. 2). New York, New York: Institute of Electrical and Electronics Engineers.

Royds Garden Limited and Worley Consultants, 1985. *Stewart Island Electricity Supply.* Invercargill, New Zealand: Southland Electric Power Supply.

Salter, S. H., 1988. *Alternative Energy Sources - 16th Report of the House of Lords Select Committee on the European Communities.* London, United Kingdom: Her Majesty's Stationery Office.

SEASUN Power Systems, 1991. *Ocean Energy Technology Information Module.* San Ramon, California: Pacific Gas and Electric Company, Department of Research and Development, 007.6-91.4.

SEASUN Power Systems, 1992. *Wave Energy Resource and Economic Assessment for the State of Hawaii.* Honolulu, Hawaii: State of Hawaii, Department of Business and Economic Development, Energy Division.

Seymour, R.J., 1983. *Extreme Waves in California During Winter 1983.* Sacramento, California: State of California, Department of Boating and Waterways.

Shugar, D.S., 1990. Photovoltaics in the utility distribution system: the evaluation of system and distributed benefits. In *Proceedings of the 21st IEEE PV Specialists Conference,* (in press).

State of Hawaii, 1988. *The State of Hawaii Data Book, 1988 - A Statistical Abstract.* Honolulu, Hawaii: Department of Business and Economic Development.

Steen J., 1985. Norwegian project: energy from the waves in the South Pacific. *Scandinavian Energy,* No. 1.

Svensson G., 1985. Swedish idea for wave power. *Energy Ahead 85,* pp. 40-44. Stockholm, Sweden: Vattenfall, Annual Research and Development Report.

Taylor, R., 1982. The availability, reliability, and maintenance aspects of wave energy. In *Proceedings of the Second International Symposium on Wave Energy Utilization,* edited by H. Berge, pp. 117-136. Trondheim, Norway: Tapir Publishers.

Thorpe, Tom, 1992. Telephone conversation with author, 21 January.

Tommerbakke, Andreas, 1988a. Telephone conversation with author, 27 August.

Tommerbakke, Andreas, 1988b. Letter to author, 3 October.

Vega, L.A., 1992. Economics of ocean thermal energy conversion (OTEC). In *Ocean Energy Recovery: The State of the Art*, edited by R.J. Seymour. New York, New York: American Society of Civil Engineers.

Versluis, J., 1980. Exposed location single buoy mooring. In *Proceedings of the 12th Annual Offshore Technology Conference*, Paper No. 3805.

Watabe, Tomiji, 1990. Telephone conversation with author, 8 November.

Watabe, T., and H. Kondo, 1989. Hydraulic technology and utilization of ocean wave power. *JHPS International Symposium on Fluid Power - Tokyo, March 1989*, pp. 301-308. Alexandria, Virginia, SEASUN Power Systems (reprint on file).

Watabe, T., H. Kondo, and M. Kobiyama, 1989. A case study on the utilization of ocean wave energy for fish farming in Hokkaido. In *Proceedings, 1989 International Symposium on Cold Regions Heat Transfer*, pp. 159-164. Alexandria, Virginia: SEASUN Power Systems (reprint on file).

YARD, Ltd., 1989. *Electricity Producing Renewable Energy Technologies Common Costing Methodology Initial Development Study, Final Report, Volume 2.* Oxfordshire, United Kingdom: Energy Technology Support Unit, Harwell Laboratory, ETSU GEN 2006 - P2.

Zangrando, Frederica, 1992. Letter to author, 6 March.

**Appendix**

The energy costing method used in this chapter computes the levelized annual revenue requirement over the book life of a power plant, and divides this by the plant's annual output. If all electricity generated by the plant were sold for this amount, the total collected revenue would have the same present value as the sum of all fixed charges and expenses paid out during the life of the plant. Levelization thus makes it possible to compare investment alternatives in terms of a single cost.

It should be noted that depreciation of the plant and periodic reinvestment in new equipment will cause the actual revenue requirement to change from year to year, whereas the levelized revenue requirement represents a constant annual payment and does not indicate the effect that a particular investment will have on actual cash flow.

Energy costs can be computed in both constant dollars and current dollars. Constant dollars were used, because they have a purchasing power more akin to the reader's recent experience. Current dollars represent what the energy would actually cost at the time payments are made.

Revenue requirements accounted for by this method are listed below. A more detailed description, including an explanation of levelization, can be found in Volume 3 of the EPRI Technical Assessment Guide (Electric Power Research Institute, 1987).

The annual revenue that must be collected to pay for the 'construction and operation of a power plant consists of fixed charges and expenses. Fixed charges are long-term financial obligations associated with building the plant and periodically replacing equipment that may wear out during the plant's life. They are "fixed" in the sense that they must be paid, regardless of how much electricity the plant generates. The following fixed charges were applied to the computation of energy costs presented in this chapter:

- Return on capital investment, in the form of debt return to creditors and equity return to shareholders.
- Book depreciation.
- Property taxes and insurance.
- Income taxes. Depreciation, property taxes, insurance, and interest on debt are assumed to be tax-deductible.

Expenses are annual payments associated with operating and maintaining the plant, and include fuel costs for diesel and coal generation. Note that the sale of co-products, such as fresh water, is credited towards expenses. If co-product sales exceed expenses, then the difference (which represents profit) is subject to income tax. For this chapter, it was assumed that fuel and other annual expenses would not be subject to real price escalation over the life of the plant.

**Financial Parameters**

The financial parameters used to compute levelized energy costs in this chapter were taken from Volume 1, Appendix A, of the EPRI Technical Assessment Guide (Electric Power Research Institute, 1986). They are as follows:

- Constant-dollar cost of debt financing:                4.6%
- Constant-dollar cost of equity financing:              7.65%
- Debt-to-equity ratio:                                  50%
- Annual inflation rate:                                 6%

  (The above combination gives a current-dollar discount rate of 12.5%)

- Federal and state income tax rate:                     38%
- Annual property tax and insurance rate:                2%

In addition, straight-line depreciation was assumed over the entire project book life, with zero net end-of-life salvage value (salvage pays for decommissioning). It was also assumed that there would be no tax preferences or credits.

**Fabrication Costs for Absorber/Reaction Structures**

Where identifiable, absorber/reaction fabrication costs were subtracted and replaced with costs thought to be representative of small fabrication yards in the United States. The original and modified cost estimates are as follows:

| Project | Structure | Original Unit Cost | Assumed Unit Cost |
|---------|-----------|--------------------|--------------------|
| Pivoting Flap | concrete caisson | 85,700 ¥/tonne | US$ 1,000/m$^3$ |
| Circular SEA Clam | concrete spine | unknown | US$ 1,000/m$^3$ |
| Danish Heaving Buoy | ferrocement buoy | 3,310 DKR per m$^2$ of deck area | US$ 300 per m$^2$ of deck area |
| Danish Heaving Buoy | deadweight anchor (concrete) | 4,115 DKR/m$^3$ | US$ 1,000/m$^3$ |

For the Pivoting Flap device, caisson weight was originally given as 70 tonnes per meter of breakwater length (Watabe, Kondo, and Kobiyama, 1989). Weight was converted to reinforced concrete volume by assuming a density of 2.61 tonnes per cubic meter. It should also be noted that both Swedish heaving buoy designs use the same unit fabrication cost for ferrocement buoys (US$ 300/m$^2$).

## Currency Exchange and Inflation

Cost estimates originating outside the United States were converted to U.S. dollars (US$) using the exchange rate in effect at the time the original cost estimate was made. Exchange rates were obtained from the International Monetary Fund in Washington, DC, and are as follows:

| Table | Project | Original Currency | Exchange Rate |
|---|---|---|---|
| 1 | Tapered Channel (Toftestallen) | Norwegian crown (NOK) 1985 annual average | 8.60 NOK/US$ |
| 1, 4 | Tapered Channel (King Island) | Norwegian crown (NOK) September 1990 average | 6.07 NOK/US$ |
| 2 | Double-Acting MOWC (Toftestallen) | Norwegian crown (NOK) 1985 annual average | 8.60 NOK/US$ |
| 2 | Single-Acting OWC (Kashima Port) | Japanese yen (¥) 1989 annual average | 138 ¥/US$ |
| 2 | Pivoting Flap (Muroran Port) | Japanese yen (¥) 1989 annual average | 138 ¥/US$ |
| 3, 7 | Circular SEA Clam (Lewis) | British pound (£) Mid-year 1989 | 1.61 US$/£ |
| 3 | Danish Heaving Buoy (Hanstholm) | Danish crown (DKR) June 1988 average | 6.67 DKR/US$ |
| 4 | Hydro and Diesel (Stewart Island) | New Zealand dollar (NZ$) 1985 annual average | 0.498 US$/NZ$ |
| 7 | Tidal Power (Callanish) | British pound (£) 1985 annual average | 1.30 US$/£ |
| 7 | Tidal Power (Langstone Harbor) | British pound (£) 1986 annual average | 1.47 US$/£ |
| 7 | Tidal Power (Severn Estuary) | British pound (£) Mid-year 1989 | 1.61 US$/£ |

Original cost estimates made prior to 1990 were escalated at an annual inflation rate of 6%. Currency exchange, where necessary, was made before escalation. Note that cost data for the Tapered Channel plant on Java (Table 1) and the MOWC plant on Tongatapu (Table 2) were originally in U.S. dollars.

## 10:  State of the Art in Other Ocean Energy Sources

Richard J. Seymour, Member ASCE[1]
Preston Lowrey[2]

Abstract

Less mature technologies are reviewed. The potential for increased extraction of ocean water constituents eventually useful in fusion power production, including lithium and deuterium, is discussed. A series of research programs conducted in the late 1970s and early 1980s are described in which macroalgea (kelp) was evaluated as biomass for generating methane as a natural gas substitute. The concepts for extracting energy from open ocean currents are detailed with additional treatment of related technology from run-of-the-river turbines. The status of wind energy extraction over water is described. The use of salinity differences between fluids is discussed for direct energy generation and for the production of fresh water from saline sources. A potential method for direct extraction of hydrogen from sea water through solar radiation is noted.

Introduction

The preceding chapters in this book have dealt with technologies for renewable energy extraction from the ocean which have already achieved a significant level of development. A number of other potential technologies exist, ranging in maturity from concepts to laboratory experiments and field investigations. This chapter will discuss the state of the art in these less-developed technologies. The more mature energy extraction methods already discussed depend solely upon mechanical or thermal resources in the ocean, while some of the systems treated in this chapter are based upon oceanic chemical resources.

Extraction of Fuels for Nuclear Fission Energy:

Every proposed cycle for fusion energy production utilizes some combination of lithium, deuterium or tritium (Isaacs and Seymour, 1973). Both lithium (fifteenth most common element in seawater at about 425 ppm) and deuterium (about 160 ppm) are readily available in the ocean. Tritium would most likely be bred in a lithium blanket surrounding a reactor. Broad commercialization of fusion power would probably require the development of a large scale ocean extraction industry for either lithium or deuterium, or both. The $Li^6$ isotope comprises about 7% of natural lithium and 1 kg has been estimated to have an energy equivalent of 60 barrels of petroleum. In addition, metallic lithium is widely used as anodes for high energy storage batteries. Epstein et al. (1981) describe a process for extracting lithium from Dead Sea brines. Solar ponds or fresh water distillation processes can also be used as the brine concentration mechanism. Aluminum chloride (recovered later in the process) is used to

---

[1]Head, Ocean Engineering Research Group, Scripps Institution of Oceanography, University of California San Diego, La Jolla, CA 92093-0222.
[2]Assoc. Professor, Dept. of Mechanical Engineering, San Diego State University, San Diego, CA, 92182-0191.

precipitate the lithium as an oxide, mixed with aluminum oxide. After separation by conventional means, the solids are dissolved in hydrochloric acid and lithium chloride is extracted by means of a water-immiscible alcohol solvent. The production of metallic lithium from its chloride by electrolysis in a fused-salt bath is an established process. Epstein *et al.* (1981) estimated that the production costs at an annual rate of 45 tonnes (about 1.25% of the world demand) would be equal to the 1980 selling price ($30/kg). Therefore, some major increase in the world demand, such as might be expected from the commercialization of fusion power, would be required to make this approach competitive with terrestrial supplies.

Deuterium is a promising fuel for fusion power. Significantly enhanced concentrations of deuterium in fresh product water can be achieved through proper selection of the distillation process (Shimizu and Wada, 1978). The conventional process for extracting heavy water (deuterium) utilized hydrogen sulfide and resulted in a toxic and corrosive effluent. Current technology, utilizing hydrogen and a hydrophobic platinum catalyst (Rolston *et al.*,1975) allows the deuterium-depleted water to be safely used for other purposes. As in the case of lithium, the evaporation of sea water to produce fresh water, as in an open cycle OTEC plant, can provide an economical enrichment step. However, in the case of the lithium, it is enriched in the brine, while the deuterium enrichment occurs in the fresh water.

Marine Biomass Production:

The discussion of marine biomass will be limited in this chapter to at-sea production. Biomass cultivation of marine species on land, as an adjunct to an OTEC plant for example, is discussed in Chapter 15.

Marine macroalgae are commercially cultivated in the Orient, but only for food. The dominant species is *Laminaria japonica*. About 1.3 million tonnes (wet weight) was produced in 1979 in China on about 18,000 hectares of farms in protected waters. In Japan, the total is far less - about 7.5 thousand tonnes - because of the very labor-intensive nature of the cultivation (Tseng, 1981). For comparison, the highly mechanical harvest of natural kelp in Southern California, largely for food additives, is about 150 thousand wet tonnes per year (OTA,1980). Diseases and other problems reported in the Orient include toxic bacteria, herbivorous crustaceans and encrustation by colonizing animals (Tseng, 1981).

The concept of culturing fast-growing macroalgae in the ocean as a feedstock for methane production is attributed to H.A. Wilcox in the early 1970s [e.g., Wilcox (1977).] In the period from 1976 through 1983, The Gas Research Institute sponsored a research and development program aimed at a system for culturing giant kelp (*Macrocystis pyrifera*) off the Southern California coast with on-shore processing to produce methane and with ocean disposal of the residue from the extraction process. This program involved studies of the growth and nutrition of the kelp, the hardware involved in deep water farms, and the chemical processes involved with anaerobic digestion of the kelp to produce methane. The discussion here will be limited to the ocean engineering technology associated with this concept.

Because giant kelp grows from a root-like holdfast in water depths of only a few tens of meters, and because tens of thousands of hectares of kelp farm would be required for a viable operation, it is necessary to consider subsurface platforms or frames to support the plants in deep water, far from shore. To sustain the high growth rates (as much as 50 cm/day) that make this kelp attractive as a biomass source, heavy fertilization with nutrient rich bottom water is required. In nature, in shallow water, this occurs through natural wind-driven upwelling along the coasts. In deep water, this upwelling must be achieved through pumping. To achieve growth rates comparable to those that occur naturally in shallow water, pumped upwelling rates equivalent to almost a meter per day are required. This follows the convention used in terrestrial irrigation of describing the amount of water added as though it were deposited at an even depth on the surface. The energy requirements for this pumping greatly exceeds (perhaps by a factor of five) the energy content of the methane produced from the predicted kelp output. This apparently nonsensical result must be understood in light of a real concern in the early 1980's that natural gas supplies would be depleted much before those of petroleum. Since that time, of course, natural gas reserves have been very substantially increased to the point where there is literally a glut of this fuel in many parts of the world. The energy to pump the necessary nutrients was proposed to be supplied from a renewable source through the use of wave-powered pumps distributed within the kelp farm. These pumps were not developed as part of this study.

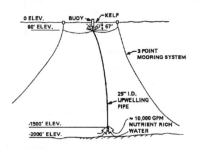

Figure 10.1: The arrangement of a proposed research kelp farm element in the California Program. [After Tompkins and Bryce (1984)]

The drag of ocean currents on the floating equipment with many millions of huge kelp plants would greatly exceed the state of the art in mooring systems and require some technology development. The most critical structural problems encountered, however, were with the plants themselves. Even in shallow coastal waters, where the water depth limits the maximum wave height, wave forces in severe storms can remove almost all of the mature plants (Seymour et al., 1989.) In the unprotected waters offshore, the problem is magnified. Further, the huge kelp wrack resulting from farms that are tens of kilometers on a side from such a storm could result in unacceptable environmental impacts. If the material sank, the resultant oxygen demand as it decomposed would render large areas of the ocean floor anaerobic and destroy existing bottom communities. If a substantial amount floated, buoyed by its gas bladders, it would come ashore in quantities that would overwhelm shoreline communities - especially those involving human use of beach recreation facilities.

With the destruction of all of the at-sea experimental facilities in the winter storms of 1982-83, and the realization that natural gas supplies were not in immediate jeopardy, this program was abandoned in 1983 and the emphasis for biomass returned to harvesting on land. The final report of the system contractor is found in Tompkins and Bryce (1984). An economic analysis indicated a cost of $12.80 per million kJ in 1983 dollars, without by-product retrieval. With by-product utilization, the costs were estimated to drop by from $1-2/million kJ.

A contemporaneous program was initiated in New York State in 1979, funded by a consortium of federal, state and industrial agencies, as a Sea Grant project. In this program, the emphasis was on using only natural processes for growing local species of macroalgae on man-made substrate. The plant selected was brown kelp (*Laminaria saccharina*), indigenous to the Atlantic off Long Island where the field work was performed.

The original concepts for providing substrate included such high technology solutions as a rigid structure, hexagonal in planform for nesting, which provided about 0.4 hectares of growing space (see Figure 10.2) and a jackup frame for adjusting the depth of the substrate. An international conference on seaweed raft and farm design was held in 1982 (McKay, 1983). Partially as the result of input from Chinese practitioners, the decision was made to proceed with an oriental-style system consisting mostly of wire and polypropylene ropes. As in the California program, measurements were made on drag forces associated with ocean currents acting on growing plants. Biofouling was a severe problem. Rope of approximately 10mm diameter increased in wet weight by about 0.3 kg/m/month at the test site and it was estimated that hand-brushing at monthly intervals would be required to insure effective operation.

A reasonably complete description of a number of the then current concepts for providing the marine farm structures and for upwelling water are contained in OTA (1980).

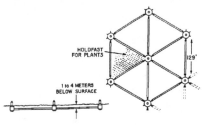

A report prepared in 1983 on the status in Europe of energy from marine biomass was unable to identify any directly relevant research supported by European national governments. This same report referred to unconfirmed rumors of a major pilot-scale marine biomass energy farm project in Japan begun in 1981 (Kayes, 1983). However, progress reports beginning in 1983 [e.g., Mitsuo *et al.* (1986)] clearly indicate that this project was intended to enhance fisheries and there was no intent to harvest the seaweed.

Figure 10.2 The proposed design of one hexagonal element for the New York Program. The frame was designed to provide one acre (0.41 hectares) of growing space. No artificial upwelling was planned. [After McKay (1984)]

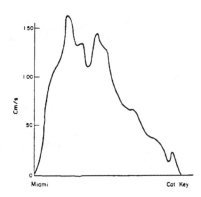

Figure 10.3: Profile of the Florida Current surface velocity near Miami, Florida. [After Lodhi (1988)]

## Energy from Ocean Currents:

Major ocean currents, powered by the trade winds in the tropics, are concentrated in poleward return flows by geostrophic forces on the western edges of the major ocean basins (eastern coasts of the continents.) The Kuroshio Current off Japan and the Gulf Stream are well recognized examples, but the Brazil, East Australian and Augulhas (southeast Africa) currents also operate in the Southern Hemisphere. The concept of extracting substantial quantities of energy from ocean currents with large turbines is attributed to W.J. Moulton in 1973 (Lissaman, 1979). A review of ocean currents and their potential energy significance is given in Stewart and Wick (1981) and in Lodhi (1988). As a result of Mouton's suggestions, the Florida Current, as the Gulf Stream is known during its closest approach to mainland USA near Miami, received considerable attention during the 1970s as a renewable energy source and was the subject of the MacArthur Workshop in March 1974 (Stewart, 1974).

This review led to the formation of a private firm, Hydro-Energy Associates, in association with Mouton and D.F. Thompson to explore the development of a Florida Current power scheme. In 1977, the team was augmented by AeroVironment, Inc., where the development effort was headed by P.B.S. Lissaman. Funding support was also provided by the U.S. Department of Energy. The R&D continued until 1981 and culminated in the design of a very large (91m) turbine of unconventional design (see Figure 10.4.) This unit utilized a slotted shroud for increased efficiency, enclosing counter-rotating turbines of about 65m in diameter. The blades in these turbines were supported between rigid rims and a central hub, similar to spokes in a bicycle wheel. The blades consisted on segments supported by a central cable which took the shape of a catenary. A gear drive on the rims powered six AC induction generators set

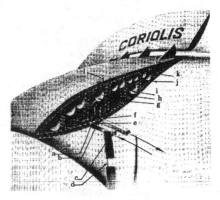

Figure 10.4: The prototype *Coriolis* design. a) generator, b) gear box, c) segmented blade, d) cable, e) friction drive, f) thrust bearing, g) pressurized access tube, h) free flooded hull, i) ballast tanks, j) aluminum plate skin, k) space frame truss, l) man for scale. After Lissaman (1979).

within the surrounding shroud. The shroud provided structural support and buoyancy for the whole turbine. The turbines were intended to be moored at a depth below the keels of ships as shown in Figure 10.5 (Radkey *et al.* (1981). The Florida Current extends to this depth with little diminution of its surface strength so that very little is lost in removing any hazard to navigation.

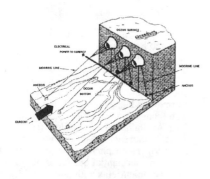

Figure 10.5: Mooring scheme for *Coriolis* turbines. After Lissaman (1979).

The turbine concept is called Coriolis, after the force (resulting from the earth's rotation) that causes intensification of western boundary currents. Each turbine was rated at 7.5 MW at a speed of the Florida Current of 1.56 m/s. Twelve such units would be moored in the same area and transmit power to shore over a single cable. Economic analyses indicated an installed cost of $2850/kW in 1981 dollars (Radkey *et al.*, 1981). As is often the case, the cost analyses appear to ignore the costs associated with frequent replacement because of fatigue on mooring and electrical cables.

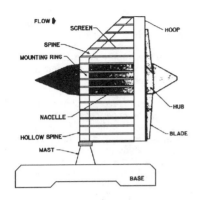

Figure 10.6: The bidirectional (reversing) KHECS 30 kW run of the river turbine. [After Miller *et al.* (1985)]

Among the technologies potentially available for ocean current energy extraction is the open turbine (run of the river turbine) proposed for use without impoundment behind dams in swiftly flowing rivers and in similar tidal flows. An example is the Kinetic Hydro Energy Conversion System (KHECS) under development by New York University and the New York Power Authority (Miller *et al.*, 1985.) This unit achieves bi-directional capability by rotating the entire turbine about a yaw bearing, as shown in Figure 10.6. The rotor uses three blades of fixed pitch, following the design of Glauert (1934), with a tip speed ratio of four. The prototype design has a rotor diameter of 4.3m, is rated at 30 kW with a 2 m/s flow, and is expected to produce an overall efficiency ("water to wire") of 0.35. The estimates of installed cost for a production run of 100 units to be installed at 10 different sites was $1700 per kW (1983 dollars.)

Wind Energy in the Ocean:

Schmitt (1981) presents an evaluation of the wind energy resource over the world's oceans and a brief review of the economics and the technology available in the late 1970s.

Heronemus (1972) appears to be the first discussion in detail of extracting energy from the wind with windmills sited at sea. This study, recognizing the problems inherent in the variability of the resource, proposed a storage system in which electrical energy would be used to electrolize sea water and the hydrogen would be then stored under ambient pressure in submerged tanks. This concept was suggested independently by Isaacs and Seymour (1973). Heronemus's paper also includes a detailed economic analysis of a 38 GW windmill farm (18 GW net output after electrolysis, hydrogen transmission to shore and conversion in fuel cells.) In 1972 dollars, the derived costs were about $0.03 kWh. Based upon the substantial body of experience with fixed and moored offshore structures that has accumulated over the

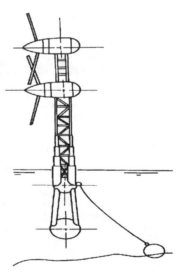

Figure 10.7: A three wheel floating wind turbine as proposed by Heronemus (1972).

past two decades, these costs seem optimistic in the extreme. As shown in Figure 10.7, Heronemus's concepts for floating towers are clearly unstable. He claimed that they would be stabilized from wave loading, functioning as a spar buoy similar to the research vessel FLIP. However, FLIP operates with more than 70% of its length submerged, and with negligible wind loads in comparison to the three large turbines. The very difficult problems of connecting a moored structure with a high voltage cable, often ignored in the preliminary evaluation of ocean power concepts, are detailed in Marine Board (1982). Similarly, Heronemus assumed that fixed towers could be simply jetted into sandy soils on shoals in the North Atlantic. Contemporary experience with wave scour at much greater depths points out the need for substantial foundations.

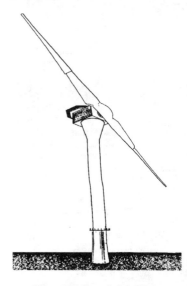

Figure 10.8: Concept for offshore installation of 100m diameter wind turbine by Burton and Roberts (1985).

A more recent design approach, which incorporates experience with land-based wind farms and with offshore platform construction, is contained in Burton and Roberts (1985). This is a study of the preliminary design and economic analysis for an array of 100m diameter wind turbines mounted on fixed offshore towers. A spacing of 10 fan diameters was chosen such that 320 machines could be sited in a 17 km square area off the East Coast of Great Britain. Each turbine is rated at 6 MW, resulting in approximately 2 GW for the array. The 10D spacing results in an estimated cluster efficiency of about 84%. This increases to 94% with a spacing of 15D.

The tower configuration selected is a gravity-stabilized reinforced concrete structure weighing 2300 tonnes. It would require a special lift barge for transportation and installation and a jackup crane rig for installing the rotor. Foundation problems with gravity based structures in the North Sea that have occurred since this study might result in the selection today of a pile base structure. The pile base was estimated to be only slightly more expensive in the analysis. The overall configuration is shown in Figure 10.8. The economic analysis bracketed the energy costs at 5 to 9 p/kWh, based upon 1984 British pounds.

## The Potential in Salinity Differences:

Salinity power refers to a large unexploited energy source which exists at the interface between water bodies with different salinities. For example, when fresh river water mixes with the sea the osmotic pressure P is 24 atm., equivalent to a 240 m head of fresh water. Between fresh water and Dead Sea brine there is 500 atm. or a 5000 m head difference (Wick, 1978). Each ft$^3$/s of fresh water flowing into the Great Salt Lake could theoretically generate 1 MW of power. Moreover, the worldwide fresh/sea water salinity resource has been estimated at 2.6x10$^6$ MW (Wick and Schmitt, 1977). The potential using saturated brines at terminal hypersaline lakes, or in underground salt deposits or solutions (Wick and Isaacs, 1976, 1978) is harder to estimate but is very large. Moreover, there are many arid shore locations where man-made hypersaline lakes or more sophisticated brine re-concentration systems could be constructed and used with seawater as the low salinity solution. Both the salinity resource size and its energy density (head) exceed the values for every other non-nuclear ocean energy resource. Most forms of salinity power depend on evaporation to maintain the salinity difference; therefore, this is generally a renewable form of energy. Moreover, usually it will be a shoreline rather than a blue water resource. This may make viable smaller power modules than is the case, for example, with OTEC. Limited attention has been directed to salinity power because it is so easy to overlook; there is no dramatic change in temperature at a river mouth, and an estuary seems far more tranquil than a 240 m waterfall.

Since the 1950s, but primarily between 1973 and 1983, a series of conceptual papers proposed various methods of harnessing salinity power (Wick, 1978, 1981). In theory any strategy that can be harnessed for desalination could be reversed to produce power, and three strategies have been seriously explored, osmosis, electrodialysis and vapor pressure differences. When exploited to produce power these strategies have been named, pressure retarded osmosis (PRO) (Loeb, 1975), reversed electrodialysis (RED) (O'Brien et al., 1984), and reversed vapor compression. To be complete, there have also been proposals to harness salinity power using electrochemical reactions (Wick and Isaacs, 1976), fibers that expand and contract as salinity varies (Steinberg et al., 1966, Sussman and Katchalsky, 1970, Emren, 1979), and the reverse of freezing desalination (Isaacs and Schmitt, 1980). These approaches, however, are less promising and received less investigation.

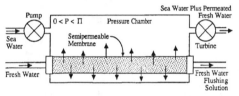

Figure 10.9: Schematic diagram of a pressure retarded osmosis (PRO) energy conversion device (Loeb, 1976, Wick, 1978). The seawater is pumped to a pressure, P, which is less than the osmotic pressure difference, Π.

Shortly after the 1973 energy crisis, the osmotic strategy received the most attention. Excitement was generated by the large osmotic pressures and equivalent hydrostatic heads cited earlier, and the seeming similarity to the mature and economical technology of hydropower. It should be remembered, however, that the salinity potential is not in this easily harnessed form of pressurized water. Rather it must be converted to pressurized water, by passing the water through selective membranes against a large fraction of the available osmotic pressure (see Figure 10.9).

In response to a seminal article by Norman (1974), Loeb (1975) realized that the costs of membranes/kW were prohibitive for a fresh water/seawater system. Loeb also realized, however, that membrane costs should drop as the inverse square of the available osmotic pressure. As a result, he projected that the osmotic salinity power strategy might already be viable using saturated brines available at the Dead Sea. A more detailed analysis (Loeb, 1976, Loeb et al., 1976, Wick, 1978), however, showed problems. The Dead Sea brines oxidized the conventional semipermeable membranes and at pressures above 100 atm. compacted the membranes and reduced their water transfer rates. As a compromise, lower operating pressures were used and the water was pre-treated. Costs were finally estimated at two to four times retail electric rates. In this period there were also proposals to install "hydro-electric-osmotic power plants" underground or in deep water offshore (Reali, 1980, 1981), however, it would probably be less costly and more environmentally acceptable to use a pressure vessel and a pump as proposed by Loeb (1976).

The immediate problem with the osmotic membranes was that membranes optimized for desalination were used in a very different environment. Little was known about the behavior of membranes in this high pressure environment with different salts.

Wick (1978) suggested possibilities that might revive the PRO strategy, such as use of new membranes with 10 times the flow rate, use of glass membranes and use of mass produced membranes to reduce costs.    This type of PRO research-- returning to membrane fundamentals-- has apparently not been undertaken in the interim.  The effect on PRO of more than a decade of progress in reverse osmosis desalination probably deserves review.

Reverse electrodialysis (RED) was the first salinity power strategy proposed and tested (Pattle, 1954a, 1954b, 1955) but remained academic until after the 1973 energy crisis.  This approach directly generates electricity from fresh and saline water using simple concentration cells.  To minimize the need for electrodes and simultaneously compound voltages the cells are arrayed in long series separated by ion exchange membranes (see Figure 10.10).  Two kinds of membranes alternate between cells; one passes anions and one cations.  The membranes must be slightly separated by spacers, and sealed into isolated compartments.  Emren (1980) contributed and tested the key idea that practical RED stacks should operate with continuous cross-flows.  This simplifies manifolding of the flows and improves performance much as cross-flow does for a heat exchanger.  Emren's group determined that electrical short circuiting in the direction of flow would not be a problem; there is sufficient electrical resistance because the cross section perpendicular to the flow is small and the flow length relatively long.  Based on the existing desalination technology, early estimates of the costs for electrodialysis membranes were also prohibitive (Wick, 1978).    Three research teams, however, independently projected costs based on mass production of improved membranes that would allow competitive production of electricity (Weinstein and Leitz, 1976, Forgacs and O'Brien, 1979, Emren, 1980).  Moreover, in contrast to the case with PRO, there have been large advances in RED membrane technology and RED cost reductions since 1978.

First, Forgacs (1978) developed polyethylene-based, very low electrical resistance, electrodialysis membranes that retained good selectivity to the desired ions (permselectivity). This led to a continuous three month field test of a RED stack which showed no evidence of membrane deterioration (O'Brien et al., 1984). Concurrently Emren and Bergström's (1982) research group in Sweden ran a seven month laboratory test on a 360 cell RED stack. This work identified a series of practical problems all tied to the anisotropic swelling of the membranes which depends on salinity. Because of this swelling, the gaps between successive

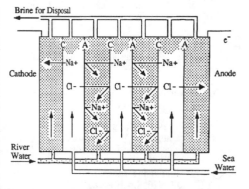

Figure 10.10: Schematic drawing of reverse electrodialysis stack (Wick, 1978). Only a few cells are shown here, but many more would be included. The A and C refer to anion- and cation-permeable membranes.

membranes became uneven. In turn, this produced uneven water flow rates which reduced performance.  Swelling also opened leaks between successive compartments through the frame and gasket.  It became obvious that simple cost lowering improvements and high quality in assembly were needed as much as sophisticated advances in membrane technology.  Although Emren proposed producing membranes with built in spacers, O'Brien (1986) has really developed this approach.  His group chemically treats ribbed polyethylene stock and converts it into ion selective membranes.  When these membranes are stacked the ribs act as self-spacers.  The ribs on successive membranes in a stack are oriented at 90° producing the cross flow channel pattern. Moreover, the edges of O'Brien's ribbed polyethylene stock retain the heat sealing property of polyethylene and successive membranes in a stack are permanently heat sealed producing leak-proof compartments. These two innovations, self-spacing and heat sealing, have eliminated leaks and the need for frames, gaskets and spacers.

Another important technical development has been the elimination of the need for sacrificial electrodes in the terminal cells in a stack. Instead, in these cells electrolysis can convert water to $H_2$ and $O_2$ or $Cl_2$ gas. The gases can later be catalytically recombined. The 2.0-3.0 volts needed will be a small loss in a big stack. Alternative terminal cell reactions that take less voltage are under investigation (Emren and Bergström, 1982).

Emren (1980) points out how quickly membranes have improved recently. In five years power density rose by a factor of seven, areal electrical resistance fell by 40 times, and costs dropped by a factor of 4 to 10. He cautions, however that further improvements will come more slowly. More recently, O'Brien (1986) pointed out that membranes with greater than 90% permselectivity and very low electrical resistances are now manufactured in five countries.

These major cost reductions were produced by improving both manufacturing and membranes. Emren (1980) projected costs using then available membranes at $3-4/kW-h. Using the best new membranes he projected costs of $0.17-0.28/kW-h. With mass production of the best membranes with built-in spacers Emren projected costs of $0.05/kW-h for seawater/fresh water and only $0.013/kW-h for brine/seawater. At about the same time Forgacs and O'Brien (1979) projected costs of $0.10/kW-h for seawater/fresh water and only $0.02-0.05/kW-h for brine/seawater. Both Emren (1980) and O'Brien (1986) now believe the technology is approaching the point where pilot plants will be seriously considered. In Canada, a private company, RE Tech Ltd., has formed to pursue development. The first installation would probably use brine/brackish water since brackish water gives a lower stack resistance than completely fresh water, and the larger salinity difference gives lower costs. RED stacks can be very modular and availability will therefore be high.

More research appears necessary to evaluate stable lifetimes of membranes, bio-fouling and its control, and whether new swelling related problems arise when modules are scaled up. In addition, further improvements in membrane performance, cost and assembly will be welcome.

In the late 1970's, the apparent limitation for both the osmotic and electrodialysis strategies was the huge capital cost for their membranes. Consequently, attention turned to the vapor pressure difference strategy (Wick and Isaacs, 1976) because it avoids all membranes. In particular, effort concentrated on reversing the vapor compression method of desalination. This approach uses two interconnected chambers, one containing fresh water and one containing brine. A vacuum pump removes air and non-condensibles from both chambers. When they start at the same temperature, the vapor pressure over the fresh water will be higher than the vapor pressure over the brine which will drive vapor through a turbine placed in the interconnecting line. Once through the turbine, water vapor will be absorbed into the brine and release its latent heat of vaporization. This heat would quickly elevate the temperature of the brine raising the vapor pressure and halting further vapor transfer. Moreover, vapor production over the fresh water will extract latent heat of vaporization from the fresh water reducing its temperature and vapor pressure. To avoid this, a heat exchanger must be installed to return latent heat released in the brine to the fresh water.

Olsson *et al.* (1979) made preliminary laboratory tests of this strategy.    The first design used two chambers separated by a sheet of copper which acted as the heat exchanger. This device operated in batch mode and was rocked to keep the copper coated with the two solutions.  The interconnecting line did not actually hold a turbine, however, the potential power production was estimated from the measured pressure difference and mass transfer rate.  In the best case this gave 7 $W/m^2$ of heat exchanger area.  This power density exceeded the highest values then projected for RED, 1.9 $W/m^2$ (Weinstein and Leitz, 1976).  The power density for PRO was even lower.  Moreover, heat exchangers should be cheaper than either osmotic or electrodialysis membranes.

With this encouragement, Olsson (1982) built a larger reversed vapor compression test unit.  This contained an ingenious double spiral heat exchanger coiled inside a 6 inch diameter PVC pipe which functioned as the external pressure vessel (see Figure 10.11).  The two required chambers were, therefore, both spiral in shape and each confined between the two spirals of the heat exchanger.  Ports arranged at alternate ends of this double spiral allowed the fresh water and brine to enter these two spiral chambers.  The whole PVC pipe unit was slowly rotated counter to the direction of the double spiral.  This rotation simultaneously pumped the two fluids into their chambers and spread thin coatings on the heat exchanger surfaces.  The double spiral design eliminated the need for separate pumps and manifolds while facilitating simple construction of a compact heat exchanger.  In this unit a small turbine was installed but it was not attached to a load.  A strobe showed that the turbine spun at several 1000 rpm.  Other test results were not reported for this unit.

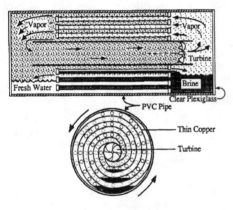

Figure 10.11: Schematic cross section (top) and end-on view (bottom) of the "double spiral" pump for converting salinity gradient energy (Olsson, 1982).

There are many similarities between this reversed vapor compression approach and both the closed (see Chp. 5) and open (see Chp. 6) OTEC cycles. Plastics, under consideration for the closed OTEC cycle, may be appropriate for the large area, low temperature difference heat exchanger needed in an Olsson type device. Large diameter low pressure steam turbines akin to helicopter blades are required in the salinity cycle as in the open OTEC cycle. Similarly, in both cycles large volumes of water must be handled and de-aerated, and a vacuum must be maintained in a large volume chamber.

If the operating temperature is elevated, the vapor pressures will go up dramatically. Consequently, more power could be extracted from a given volume module. This would not, however, change the power produced per mass of fresh water consumed. In addition, there might have to be a heat source at the elevated temperature for two reasons. First, these devices convert internal energy to work therefore, the temperature will fall unless heat is added. For a few salts, such as $CaCl_2$, the required heat could be the heat of dilution. Second, flow-through operation will require a heat exchanger inserted between a) the incoming ambient temperature fresh water and concentrated brine and b) the warm spent, diluted brine. This could scavenge most but not all of the heat in the spent brine.

For engineers experienced with conventional power cycles, harnessing this vapor pressure difference appears very difficult. At ambient temperatures, for fresh water/saturated NaCl it is equivalent to expanding steam through only 4-5°C. For fresh/seawater it is equivalent to only 0.5°C. Conventional fossil-fueled steam cycles often use 500°C and even OTEC would use 18-20°C. Terminal lakes rich in $MgCl_2$ and $CaCl_2$ may have more promise (Isshiki, 1977). For example, there is a design for a reversed vapor compression power plant at elevated temperatures for the $MgCl_2$ rich Dead Sea. This has an equivalent temperature difference of 22°C (Nadev and Ophir, 1982).

The spent $MgCl_2$ would have to be re-concentrated. It therefore seems that the evaporation rate from the Dead Sea, or similar terminal lakes, would limit the capacity of such power plants. Assaf (1984, 1986, 1987), however, has recognized that re-concentration is governed, not by the solar insolation, but by mass transfer rates to the low ambient humidity. So he has designed wind aided showers for re-concentrating diluted brines. This technology will add a capital expense, but could make salinity power based on $MgCl_2$ or $CaCl_2$ a virtually unlimited resource.

At solar salt production works, after precipitating the maximum amount of pure NaCl, the residual saturated salt solution is called bittern. Bittern is now a waste product but it is rich in $MgCl_2$. For the vapor pressure difference strategy, therefore, the contribution of the oceans may turn out to be as a source of highly absorbent bittern. Generally, this would be available in arid shore locations, and would be combined with vapor drawn from seawater.

Intuitively, the potential to produce work using brine and fresh water bodies must be the same regardless of the method used to extract it. It therefore seems incongruous that the osmotic pressures can exceed 240 atm (3528 psia) while simultaneously the vapor pressure differences are only 0.2 psi. How can these two strategies give identical amounts of work? This can be understood by examining the formula for reversible steady-flow work,

$$w = - \int_1^2 v dP,$$

where w is work per unit mass of the working fluid, v is specific volume, and P is pressure. This formula applies to both extraction methods. Near 20°C, the specific volume of low pressure steam is 4 to 5 orders of magnitude larger than the specific volume of liquid water for a hydroturbine. So to produce the same work per mass of working fluid the change in pressure in the steam should be and is 4 to 5 orders of magnitude smaller than that in the liquid water.

In the 1970s Wick and Isaacs (1976) saw the major salinity power resource as that between seawater and fresh water at the mouth of rivers. From an engineering perspective, however, it is more likely that using concentrated brine and either fresh or seawater has a better chance to become economical initially. This is because the potential difference is 9 or 10 times as large. Consequently, about one tenth the volume flow of water must be processed which greatly reduces many costs -- for pipes, pumps, pressure vessels, etc. In addition, for the PRO and vapor pressure difference strategies, costs should drop with the inverse square of the potential difference. For RED power is proportional to the square of the *ln* of the concentration ratio (Ohya, 1983). So using saturated brine drops costs by only a factor of 2-5. Also, the cost of all three strategies should go down if the ambient temperature is higher; therefore, the first installations should be in warm climates, for example arid desert shores. Lastly, the environmental costs will probably be less or at least more politically acceptable at natural or man-made terminal brine lakes than at river mouth estuaries.

Most of the investigation of salinity power has focused on generating electric power. There has also been intermittent interest in using salinity potential to directly drive desalination. This was first proposed with osmosis (Murphy, 1958). Water transferred through an osmotic membrane from a brackish solution would increase the pressure in a confined concentrated solution. In turn this pressure, acting through a piston, could force fresh water out of a second brackish solution through a second osmotic membrane. Alternatively RED cells in series (see Figure 10.12) could use seawater and brackish water to produce an electric potential that would force salt out of a separate brackish water cell leaving fresh water behind (Wick and Isaacs, 1976). This would have an added advantage; it would eliminate even the terminal electrodes needed in an electrodialysis stack. Work on this approach is reportedly continuing at the Weizman Institute in Israel (Friedman, 1989). An analogous scheme for salinity powered desalination uses the vapor pressure difference strategy rather than osmosis or electrodialysis. A temperature rise in brine develops by absorbing vapor from seawater in a vacuum. This temperature then drives a single distillation stage. This is now undergoing early tests and modeling by one of the authors (Lowrey). With vapor pressure difference cycles using $MgCl_2$ or $CaCl_2$ direct open cycle heat pumping or air-conditioning (Collier, 1979, Kakabaev, et al., 1976, Wood, et al., 1988) is also under study.

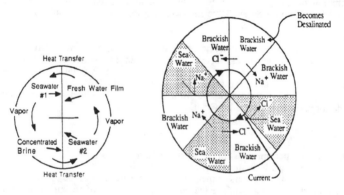

Figure 10.12: Two schemes for using salinity power to drive desalination. In one, a temperature rise produced by a salinity difference and vapor transfer in a vacuum drives a distillation stage. The second approach uses a stack of reverse electrodialysis cells to produce a current that produces electrodialysis in a separate brackish water cell.

In summary, there is a large unharnessed energy resource wherever solutions of different salinity combine. Specifically, the salinity potential of the oceans relative to the fresh water in rivers is equivalent to a hydrostatic head 240 m high, coincidentally near the average source height of the rivers of the world. Consequently, this category of salinity potential is equal to the hydroelectric potential of the world. The potential difference where low salinity waters mix with saturated brines in terminal lakes can easily be 10-20 times as great. There are three major strategies for harnessing salinity potential, using osmosis, electrodialysis or vapor pressure differences.  Each of these has been extensively developed for desalination and can be reversed, in principle, for harnessing salinity power.

Pressure retarded osmosis (PRO) is far too costly for use with fresh and seawater but its cost drops with the inverse square of the osmotic pressure difference.  Still, in the 1970s it was too costly by a factor of 2-4 when used with saturated brines.  Since then, the research to customize the membranes and make them more economical has apparently been insufficient.

Reverse electrodialysis (RED) is also governed by the cost of the membranes involved.  Major progress has been made raising the performance of membranes while lowering costs.  Also, strategies have been developed which could eliminate problems with gaskets, frames, spacers, leaks and electrodes.  The two leading researchers in the field both report that the technology is approaching the point where pilot plants should be considered.  Research remains to be done on reliable manufacturing and durability of membranes.  RED is the only salinity power strategy that has a chance to be economical when used with fresh/seawater, although it will probably be used first with brackish water/brine.  RED may also work in salinity powered desalination.

The vapor pressure difference strategy has now focused primarily on using salts other than NaCl, specifically $MgCl_2$ or $CaCl_2$.  Here, seawater would often function as the relatively fresh solution instead of as the absorbing solution.  Another role for the ocean may be as a source of $MgCl_2$ rich bittern.  Also, this type of salinity power could develop in arid coastal areas.  Salinity powered desalination driven by vapor pressure differences may have a better chance at becoming economical than electricity production.

Direct Solar Radiation:

Prof. H.T. Tien of Michigan State University has recently developed a method for direct extraction of hydrogen fron seawater using solar radiation (Worthy, 1989.)  The process emulates photosynthesis in that it employs a membrane separating two aqueous solutions. Solar radiation impinging on the membrane brings about chemical oxidation and reduction on opposite sides. The membrane used is a ceramic/metallic semiconductor and could be rugged enough for power applications.  About 10% of the solar radiation energy is converted to an equivalent energy content of hydrogen. The arrangement of the experimental apparatus is shown in Figure 10.13.

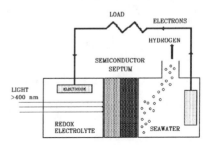

Figure 10.13: Schematic arrangement of solar cell for direct generation of hydrogen from seawater. [After Worthy (1989).]

References

Assaf, G., 1984. "Hygroscopic dew point conversion," *PCH Physiochem. Hydrodyn.* Vol. 5 (5-6) pp. 363-368.

Assaf, G., 1986. "Apparatus for producing power using concentrated brine," *U.S. Patent #4,617,800.*

Assaf, G., 1987. "Apparatus for producing power using concentrated brine," *U.S. Patent #4,704,993.*

Burton, A.L. and Roberts, S.C., 1985. "The outline design and costing of 100m diameter wind turbines in an offshore area." Proceedings of the 1985 Seventh British Wind Energy Association Conference, Oxford, 27-29 March, 1985, Garrad., A. (Ed.), Mechanical Engineering Publications, Ltd., London, pp 269-277.

Collier, R.K., 1979. "The analysis and simulation of an open cycle absorption refrigeration system," *Solar Energy*, Vol. 23, pp. 357-366.

Emren, A.T., 1979. "Salinity power production in mechanochemical engines," *Energy*, Vol. 4(3), pp. 439-449.

Emren, A.T., Dec. 1980. "Concentration cell for salinity power production, economic potential of the concentration cell," *Alternate Energy Sources 3* Proc. of the 3rd Miami Int. Conf. on Alternate Energy Sources, Vol. 4, pp. 229-241.

Emren, A.T., and Bergström, S.B., 1982. "Concentration cell for salinity power production: a seven month experiment of a concentration battery," *Alternate Energy Sources 4*, Proc. of the 4th Miami Int. Conf. on Alternate Energy Sources, Vol. 4, pp. 141-151.

Epstein, J.A., Feist, E.M., Zmora, J. and Marcus, Y., 1981. "Extraction of lithium from the Dead Sea," *Hydrometallurgy*, Vol. 6, pp. 269-275.

Friedman, R., Aug-Sept 1989. "Seawater to drink," *Technology Review,* pp. 14-15.

Forgacs, C., 1978. "Generation of electricity by reverse electrodialysis," R&D Authority, Ben Gurion University, Beer Sheva, Israel.

Forgacs, C., and O'Brien, R.N., 1979. "Utilization of membrane processes in the development of nonconventional, renewable energy sources," *Chemistry in Canada,* Vol 31(5), pp. 19-21.

Glauert, H., 1934. "Windmills and fans," In *Aerodynamic Theory*, Vol. IV, C. XI, Div. L., W.F. Durand (ed), reprinted by Peter Smith, Gloucester, MA, 1976, pp. 324-340.

Heronemus, W.E., 1972. *Proc. 8th Ann. Mar. Tech. Conf.*, Washington, DC, pp. 435-466.

Isaacs, J.D., and Schmitt, W.R., 1980. *Science*, Vol. 207, pp. 265.

Isaacs, J.D. and Seymour, R.J., 1973. "The ocean as a power resource," *International Journal of Environmental Studies*, Vol. 4, pp. 201-205.

Isshiki, N. 1977. "Study on the concentration difference energy system," *Journal of Non-Equilibrium Thermodynamics*, Vol. 2(2), pp. 88-107.

Kakabaev, A, Khandurdyev, A., Klyshchaeva, O. and Kurbanov, N., 1976. "A large-scale solar air-conditioning pilot plant and its test results," *International Chemical Engineering*, Vol. 16(1), pp. 60-64.

Kayes, R.J., 1983. "Energy from marine biomass in Europe," Political Ecology Research Group, Oxford, Rpt. RR-10, 46 pp.

Lissaman, P.B.S., 1979/80. "The Coriolis Program," *Oceanus*, Ocean Energy, 22(4), pp. 23-28.

Lodhi, M.A.K., 1988. Power potential from ocean currents for hydrogen production, *International Journal of Hydrogen Energy*, Vol. 13(3), pp. 151-172.

Loeb, S., 1975. "Osmotic power plants," *Science*, Vol. 189, pp. 654-655.

Loeb, S., 1976. "Production of energy from concentrated brines by pressure-retarded osmosis: I. Preliminary technical and economic correlations," *Journal of Membrane Science,* Vol. 1, pp. 49-63.

Loeb, S., Van Hessen, F., and Shahaf, D., 1976. "Production of energy from concentrated brines by pressure-retarded osmosis: II. Experimental results and projected energy costs," *Journal of Membrane Science,* Vol. 1, pp. 249-269.

Marine Board, 1982. *Ocean Engineering for Ocean Thermal Energy Conversion*, National Academy Press, Washington, DC, 69 pp.

McKay, L.B. (Ed.), 1983. *Seaweed raft and farm design in the United States and China.* New York Sea Grant Institute, Albany, NY, 100 pp.

McKay, L.B. (Ed.), 1984. *Marine Biomass: New York State Species and Site Studies.* New York Sea Grant Institute, Albany, NY, 72 pp.

Miller, G., Corren, D., Armstrong, P., Franceschi, J., Tan, C. and Stillman, G., 1985. "Kinetic hydro energy conversion system study, Proceedings of Waterpower '85, American Society of Civil Engineers, Las Vegas, NV, Sept. 25-27, 1985, pp. 1022-1033.

Mitsuo, N., Nazumi, T. and Fujita, M., 1986. "Experiment of developing an artificial seaweed farm plant in the Japan Sea," *Conference Record*, 14th Meeting of U.S.-Japan Marine Facilities Panel, U.S./Japan Cooperative Program in Natural Resources, Bethesda, MD, Sept. 19-20, 1986, pp. 37-40.

Murphy, G.W., 1958. "Osmoionic demineralization," *Ind. Eng. Chem.*, Vol. 50, pp. 1181.

Nadav, N. and Ophir, A., 1982. "Production of electrical power from concentrated brines," *Desalination*, Vol. 40, pp. 197-211.

Norman, R.S., 1974. "Water salination: A source of energy," *Science,* Vol. 186, pp. 350-352.

O'Brien, R.N., 1986. "Reverse electrodialysis -- an unused power source?" *Chemistry in Britain*, Vol. 22(10), pp. 927-929.

O'Brien, R.N., Visaisouk, S. and Turnham, B.D., May 1984. "Electric power from salinity gradients by reverse electrodialysis," Energy Developments: New Forms, Renewables, Conservation Conference, ENERGEX' 84, The Global Energy Forum, pp. 143-146.

Ohya, H., 1983. "Diaylitic battery covertible free energy of mixing of sea water and river water," Pacific Chemical Engineering 3rd Conference, Vol. 3, pp. 451-456.

Olsson, M.S., Wick, G., and Isaacs, J.D., 1979. "Salinity gradient power: utilizing vapor pressure differences," *Science*, Vol. 206, pp. 452-454.

Olsson, M.S., 1982. "Salinity-gradient vapor-pressure power conversion," *Energy*, Vol. 7(3), pp. 237-246.

OTA, 1980. "Fuel from open ocean kelp farms.," Office of Technology Assessment, U.S. Government Printing Office, Washington, DC.

Pattle, R.E., 1954a. "Production of electric power by mixing fresh and salt water in the hydroelectric pile," *Nature*, Vol. 174, pp. 660.

Pattle, R.E., 1954b. "Experiments on the production of electricity without fuel using the first model of the hydroelectric pile," Report to the National Research Development Corporation.

Pattle, R.E., 1955. "Electricity from fresh and salt water -- without fuel," *Chemical and Process Engineering*, Vol. 36, pp. 361.

Radkey, R.L., Mouton, W.J. and Lissaman, P.B.S., 1981. "Coriolis Program Phase III: Technical and Economic Evaluation, Final Report, Solar Energy Research Institute, Golden, Colorado, SERI/TR-9241-1, 176 pp.

Reali, M., 1980. "Closed cycle osmotic power plants for electric power production," *Energy*, Vol. 5, pp. 325-329.

Reali, M., 1981. "Submarine hydro-electric-osmotic power plants for an efficient exploitation of salinity gradients," *Energy*, Vol. 6, pp. 227-231.

Rolston, J.H., den Hartog, J. and Butler, J.P., 1975. "The deuterium isotope separation factor between hydrogen and liquid water," Atomic Energy of Canada Ltd., Chalk River, Ontario. Chalk River Nuclear Labs, March, AECL-5025, 29p.

Schmitt, W.R., 1981. "Wind energy", In: *Harvesting Ocean Energy*, Wick, G.L. and Schmitt, W.R. (eds.), UNESCO Press, Paris, pp. 150-171.

Seymour, R.J., Tegner, M.J., Dayton, P.K. and Parnell, P.E., 1989. "Storm wave induced mortality of giant kelp, *Macrocystis pyrifera*, in Southern California," *Estuarine, Coastal and Shelf Science*, Vol. 28, pp. 277-292.

Shimizu, M. and Wada, T., 1978. "Sea water desalting and heavy water production," *Proceedings* 6th International Symposium on Fresh Water from the Sea, Vol. 1, pp. 55-64.

Steinberg, I.Z., Oplatka, A. and Katchalsky, A., 1966. "Mechanochemical engines," *Nature*, Vol. 210, pp. 568-571.

Stewart, H.B., Jr. (Ed.), 1974. *Proceedings*, MacArthur Workshop on Energy from the Florida Current, Palm Beach Shores, FL, February, 1974.

Stewart, H. and Wick, G.L., 1981. "Ocean-current energy", In: *Harvesting Ocean Energy*, Wick, G.L. and Schmitt, W.R. (eds.), UNESCO Press, Paris, pp. 133-147.

Sussman, M.V., and Katchalsky, A., 1970. "Mechanochemical turbine: A new power cycle," *Science*, Vol. 167(3914), pp. 45-47.

Tompkins, A.N. and Bryce, A.J., 1984. "Marine biomass program final report," Gas Research Institute, GRI-84/0182, 166 pp.

Tseng, C.K., 1981. "Seaweed cultivation in China," In: *Biology of Seaweeds*, Lobban, C.S. and Wynne, M.J., eds. Botanical Monographs, Vol. 17, Blackwell, Oxford.

Weinstein, J.N. and Leitz, F.B., 1976. "Electric power from differences in salinity: The dialytic battery," *Science*, Vol. 191, pp. 557-559.

Wick, G.L., and Isaacs, J.D., 1976. "Utilization of the energy from salinity gradients," Institute of Marine Resources, University of California, La Jolla, CA. 92093, IMR Ref. No. 76-9, pp. 1-33.

Wick, G.L. and Schmitt, W.R., 1977. "Prospects for renewable energy from the sea." MTS Journal, Vol. 11 (5&6), pp. 16-21.

Wick, G.L., and Isaacs, J.D., 1978. "Salt domes: Is there more energy from the salt than from the oil?" *Science*, Vol. 199, pp. 1436-1437.

Wick, G.L., 1978. "Power from salinity gradients," *Energy*, Vol. 3, pp. 95-100.

Wick, G.L., 1981. "Salinity energy." In: *Harvesting Ocean Energy*, Wick, G.L. and Schmitt, W.R. (eds.), UNESCO Press, Paris, pp. 111-131.

Wilcox, H.A., 1977. "The ocean food and energy farm project (Naval Undersea Center, San Diego, CA.)," *Proceedings*, European Seminar on Biological Solar Energy Conversion Systems, Grenoble-Autrans, France, May 9-12, 1977.

Wood, B.D., Siebe, D.A. and Collier, R.K., 1988. "Open-cycle absorption system a low-cost solar heating and cooling option," *presented at the ASME winter annual meeting*, Chicago, Ill., 88-WA/Sol-10, pp. 1-12.

Worthy, W., 1989. "Solar cell produces hydrogen from seawater," *C&EN* (June 26), p. 37.

# 11: FACILITATING TECHNOLOGY FOR ELECTRIC POWER GENERATION

Ian Pope [1]

## Abstract

*This chapter is concerned with existing electric power systems, and suggests new features that would enable ocean energy sources to be incorporated into them on a significant scale. The perspective is primarily of the operation of power systems, both large and small, and how present practice, proven new technology and anticipated developments might be adopted in a satisfactory manner. Energy storage plus demand side management are envisaged as furthering the introduction of ocean energy.*

## Introduction

The oceans provide part of the abundant sources of renewable energy which, although technically available, have hardly (apart from hydro-electricity) been harnessed. Mankind needs to employ its ingenuity, firstly to reduce the costs of capturing this energy, and secondly to use it appropriately. In most instances electricity is the best way of moving energy from the point of production to the point of demand, and it is often easiest to use in an efficient manner. However, electricity suffers in comparison with other forms of energy because it is not readily stored.

It is vital to learn how best to harness renewables for electricity supply systems, for they will be needed both in developed countries as a substitute source and in developing countries as the basis for increased consumption. Proven technology already exists to obtain energy from some renewable sources in substantial quantities at costs approaching competitiveness with the best fossil-fuelled technology. Largely unproven technology has also been developed to utilise renewables on a significant scale in power systems, notwithstanding a key problem with renewables; intermittent or cyclic availability.

This chapter sets out the problems and considerations that have to be taken into account when planning the integration of an ocean energy source into an electrical supply network. Because of their predominance in the world electricity market, the bulk of the chapter deals with integration into large systems.

Problems vary depending upon the type of source and its size relative to the network: the factors affecting the integration of a tidal scheme of a few hundred megawatts into an interconnected system with tens of thousands of megawatts of maximum demand would be very different from those affecting the incorporation

---

[1] Ian Pope, Managing Director, L E Energy Limited, Blythbank, Station Road, Duns, Berwickshire TD11 3EJ, Scotland.

of small wave energy devices on to an island system with perhaps only a 5MW diesel plant.

Before discussing the incorporation of ocean energy projects into electrical supply systems, it is necessary to understand the principal characteristics of those systems and the range of alternative generation projects that are available to those who plan their expansion. The features of ocean energy which affect its integration into large and small power systems are then discussed, and finally there is a brief description of energy storage, bulk energy transmission and demand side management, facilities which will be needed if ocean energy sources are to take a more significant role in future power systems.

## Large power systems

Electricity supply is a public service for which there are many different organisational structures, and is usually administered either by national or local authorities or by publicly owned companies. Supply systems have evolved from the primitive systems created at the end of the last century to the present diversity of size, sophistication, and energy sources. These is a worldwide trend towards privatisation and competitive markets, but at the same time, the expansion of pools and contracted control of operators. There is one unifying feature inherent in all systems: electricity itself cannot be conventionally stored, which means that at all times the electrical power generated must equal consumer demand for power.

On small systems the government of the "quality" and quantities of electricity may be automated, but larger systems usually have a control engineer to regulate the production and/or consumer demand in order to minimise costs. This is done by ensuring that the cheaper plants are loaded first, and that only sufficient capacity to meet demand at the time is brought into service.

Regulation in the short term is catered for by fitting all generation units with governors to control their speed. As demand increases (provided there is enough capacity running to meet demand) the increase is met from stored mechanical inertia and the system slows down. The governors on the plants sense the slowing and respond by increasing output. Because ocean energy sources are constrained in both the timing and level of their power output, they are likely to require special attention during both the planning and operation of the power system to which they are connected.

Unlike a series of isolated, small generating systems, a grid transmission system generally enables economies of scale to be obtained unless the area to be supplied has only a low demand density, in the which case the cost of building long transmission lines, which suffer greater losses, may be a determining factor.

The reasons why the larger systems are at an advantage are because they allow for:

- sharing of spare generating capacity;

- lower capital costs per kW for generating plant;

- diversity of demand shape; and

- diversity of energy sources (the cheapest is used for base load).

Spare generating capacity is needed for security of supply. Power system managers generally aim to have at all times sufficient spare plant available to make good a power shortfall from "the largest single risk of sudden loss of generation". In the case of a number of smaller, discrete power systems, operators will each have to provide their own spare plant at significant cost, both in capital cost and the cost of operating as "spinning spare".

If the small power systems can be inter-connected the aggregate amount of spare capacity can be reduced to the largest single generation loss risk, thus saving on both capital and running cost. The "largest single risk" is the largest amount of generating capability which might be lost from the system because of one single incident. Typically this risk is considered to represent the largest generator, but it might be two generators if a credible single incident could put them out of service simultaneously (for example, substantial amount of shared auxiliary plant). It might also be the capacity of a heavily loaded intertie in the transmission system.

In a situation where customer demand is growing, new generating plant needs to be added in order to maintain a margin between maximum demand and total generating capability. In a small system the increased rate of demand may be insufficient to justify building a large, new plant with a low capital cost per kW. With the inter-connection of several smaller systems the requirement for new generation is likely to be greater, and will thus justify the construction of larger plant.

The bigger a power system is, the greater will be the variety of demand shapes of individual customers, such that the net maximum demand will be less than the sum of the individual maximum demands. Similarly, the load factor of the total system should be better than that of the constituent parts.

A small power system may have a limited selection of energy sources from which to obtain its generation. Interconnection should provide a wider range of sources, thus enabling maximum use of the most plentiful, cheap, and environmentally safe sources while the remainder are confined to a standby or peaking role. In addition, the larger ground area encompassed by an interconnected system may provide access to further energy sources (for example, hydro-electricity and coal at the pit head) which are not so easily obtained within the service area of a smaller system.

### Sources of electrical energy

In order to understand the opportunities and the difficulties in using any generating technology in a large power system, the power system planner should be aware of the advantages and disadvantages of all the technologies. The types of generating plant used in contemporary power systems are quite varied. This section indicates the main features (in so far as they affect the operation of the power system) of hydro-electric, fossil-fired and nuclear plant which are in widespread use around the world. There are many novel sources of energy emerging, some of which having already been applied on a significant scale are also mentioned. Generally these are less damaging to the environment. Some appreciation of the options is needed when examining the integration of ocean energy. Figure 11-1 shows a typical example of how a number of different energy sources may be used together to satisfy the daily consumer demand curve.

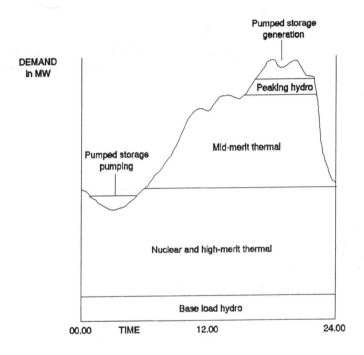

*Fig 11-1 Daily consumer demand curve satisfied from mixed thermal and hydro-electric generation and pumped storage*

## Hydro-electricity

Hydro-electricity is one of the oldest sources of energy known to man; water wheels have provided power for generations. Today, where suitable geographical conditions exist, it can provide an ideal source of electrical energy. Capital costs are generally high because of extensive civil engineering works and because of the frequent need for long transmission lines; the storage reservoir may cover a large land area. Although virtually free from emissions, hydro-electricity nevertheless often comes in for environmental criticism because of its effect on river flow regimes and on the area to be flooded. Running costs are very low and there may be other benefits from flood control and irrigation.

Hydro-electric stations at their simplest are of the "run of river" type, in which the power output depends on the flow in the river and there is no capability to store water. Thus operation occurs twenty-four hours a day, with the power output changing slowly according to the river flow. At their most complicated storage hydro-electric stations are able to store water, and thus energy, on a seasonal basis, and are able to generate at any time of the day. The only constraint is that the water level behind the dam must stay within prescribed limits.

In the median form hydro-electric stations are positioned in cascade along the course of a river, each station having some storage behind the dam creating its head. Water available at downstream stations is dependent upon the operation of upstream ones, and there may be a time delay between water leaving one station and arriving at the next. This can make management of the scheme quite complex if the output is being integrated with, for example, thermal plant. In systems where the system control engineer has the choice of using hydro-electricity, its quick start capability can be very useful. However, the energy source is not unlimited and is dependent upon the seasons and weather.

### Coal- and oil-fired generation

Coal- and oil-fired steam cycle plant is the most common source of electrical power in the world, and is usually the marginal generation used to satisfy the customer demand remaining after the basic sources have been used. Maximum thermal efficiency is limited to around 40% unless a use is found for some of the waste heat from the steam condensers. When a power system has several steam power stations they are often, for operational purposes, ranked according to their production cost, and the "merit order" is used to allocate running, the most going to the station with lowest cost production. Start up costs of such plant are high, and start up time can be as much as eight hours, although this time is reduced if the plant is still hot from a previous run, while oil-fired plant is generally quicker to start than coal-fired because of easier boiler control. Substantial quantities of fuel are stored at the power station, so generally the output should be sustainable for several weeks even if no further fuel deliveries are possible.

### Nuclear generation

Nuclear stations make a significant contribution to some power systems. Although they are generally capable of some regulation of output, economic pressures mean that they are generally classed by the system operator as "must run" plant whose output cannot be regulated. Technical complexity, very high safety costs and high capital cost mitigate against nuclear plant. Refuelling requires occasional shut down, although some gas cooled reactors are able to refuel on load. If the present research into controlled nuclear fusion were to be successful, we might eventually have access to an unlimited energy source.

### Gas turbine and combined cycle plant

Gas turbine generators were initially installed in power systems as quick start reserve plant. The construction time is comparatively short and the capital cost low which makes them useful as "quick build" capacity for systems where demand has been underestimated. However, operating cost of the aero type gas turbines is high because of the high grade fuel they require and their inherent low efficiency. More recently industrial grade gas turbines have been used, generally fired by natural gas. Their efficiency is higher than aero type gas turbines, but their response times are longer. They are now being widely used in combined cycle situations (see Figure 11-2) where the residual heat in the gas turbine exhaust is employed for steam cycle power. Efficiency is greater in industrial gas turbines and in the combined cycle mode, when over 50% efficiency can be achieved.

## Combined cycle gas turbine plant

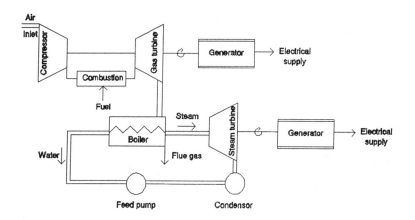

*Fig 11-2 Combined cycle generation - schematic*

Combined cycle gas turbine plant (Figure 11-2) is a recent development that in the main uses natural gas to fuel a gas turbine generator, while the hot exhaust gases pass to a steam boiler to operate a steam turbine-driven generator for additional power output. This configuration enables a thermal efficiency of over 50% to be achieved.

## Pumped storage

Pumped storage plant (see Figure 11-3) is quite widely used in large power systems. Strictly speaking it is a means of storing energy and not an energy source at all. A few percent up to 20% of the total generating capability of a power system can usefully come from this source, depending upon demand shape and plant mix. The benefits accrue from improved utilisation of steam plant (less overnight shutdowns and daytime peak lopping, and less part load running) and from the more cost effective provision of spare capacity. Pumped storage plant is not capable of sustained generation at high load levels because of the desirability of only refilling the reservoir at times when relatively cheap system energy is available.

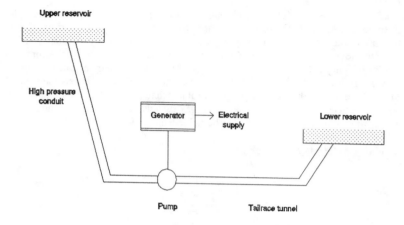

*Fig 11-3 Pumped storage - schematic*

## Renewable energy

Apart from hydro-electricity, the most promising source in many parts of the world is wind energy. The technology is being used for both small, one-off installations feeding isolated systems or the local electricity distribution systems, and for large wind farms supplying energy to the grid. For example, California has some 1400MW capacity from three main sites.

Solar radiation can be used to generate electricity by photovoltaic action or by thermal concentration; the former is already being applied to meet low power needs in remote areas.

Biomass has the potential to be a useful energy source, but its application as a means of producing electricity is likely to be on a small scale in the foreseeable future because of the vast acreage which would be needed to grow the fuel crop.

## Integration of ocean energy into large power systems

Much that has been written and spoken about the integration of ocean energy, and indeed any other type of renewable energy, into power systems is concerned with the need to store energy from "glut" periods to cover the "lean" periods. The renewable energy source then appears to power system operators to have characteristics similar to the conventional thermal plant that they are used to. There are two drawbacks to this approach: firstly, it denies a realistic assessment of the value of the "unconditioned" energy to the network; and secondly, it diverts the focus of technical development away from the reduction of construction costs and into the creation of power conditioning mechanisms and energy storage media.

Take first a simple example. Let us postulate the development of 200MW of wave energy devices off the Cornish coast in England which is fed into the UK electricity network. That network has a maximum demand of a little over 50GW (1GW is 1000MW) and a minimum demand in the middle of a Saturday night in summer of around 10GW. The grid operator is probably able to predict demand to within 200-300MW so that the presence or absence of a 200MW plant, where output is uncontrollable but predictable, is unlikely to present an insurmountable problem.

Depending upon the relationship between minimum system demand and the quantity of genuine "must run" plant, it is apparent that modest quantities of tidal energy can be taken without the need for special measures. Even the availability of tidal power in excess of the difference between minimum demand and "must run" plant need not present a problem. In this case, some tidal power would be rejected during the low level periods, but this may well only represent a minor proportion of the total potential production and thus would not affect the project economics significantly.

This basic argument can be extended to all forms of renewable inputs to a large and diverse power system. The amount of the renewable input that can be taken becomes an economic argument, for the more renewable (and presumably uncontrollable) input, the greater the probability that some will have to be rejected because of lack of demand or lack of flexibility of existing plant. The more potential energy that has to be rejected the lower the useful output to set against the initial cost, and therefore the poorer the economic performance.

Apart from considerations of the acceptability of the overall quantity three other main considerations arise:

- the predictability of the output (tidal power is predictable days in advance whereas wave energy output could probably only be forecast with some certainty a few hours ahead);

- short-term output fluctuations (the output of a single wave energy device has considerable fluctuations in output over a timescale of a few seconds, although large number of the devices would average their collective output over time to produce a smooth output); and

- the reliability of the output at peak times (different types of ocean energy device have different probabilities of producing output at times of high system demand).

The more predictable the source the better, because it allows the power system operator to adjust the other plant on the system to make the most economic use of the renewable output. There is, of course, every economic incentive to do so since the marginal cost of production at ocean energy plants is ostensibly zero. If the output rises or falls at a predictable time to predictable levels then conventional steam plant can be started and stopped accordingly, and there would be no need to increase the level of conventional generation held in reserve against the ocean energy not performing to plan.  In effect, predictability allows the ocean energy project to be credited with the full running cost of the displaced conventional plant.

Fluctuations in short term output mean that the amount of variation compensated for by the automatic governing facilities on thermal plant would have to be higher. In turn, this increases the quantity of conventional plant that has to be kept available and synchronised to the system to provide reserve.  Short-term fluctuations are only a consideration with low numbers of conversion devices, especially with a random source like wave energy.  Small numbers of devices will, by definition, be insignificant on a large power system while fluctuating output is not likely to be a problem for large systems but could be significant on small ones.  Devices with fluctuating output can be credited only with the marginal costs of reducing the output from conventional plant already connected to the network.

Reliability of output is perhaps the most difficult to balance against the other, conventional plant on the system.  Power system planners are used to modelling and evaluating the differences between plants whose output is considered to be controllable and not limited by fuel input, so it is hardly surprising that such considerations should be paramount in planners minds.    Most utilities and governments place the greatest importance on security of electricity supplies and failure to do so can result in intense public hostility.  In some countries where there is insufficient capacity and supply interruptions are commonplace, it has been known for utility staff to be physically attacked when trying to implement an interruption to prevent total system collapse.    All forms of renewable energy conversion systems present some prospect of being unavailable due to a shortage of the input energy.  There are therefore two possible failure mechanisms at time of peak demand on the network (the circumstance of most interest to power system planners):  failure of the conversion device itself;  and failure or diminution of the energy source.

Conventional plant with even a large fuel stock can fail due to breakdown. Depending on the fuel, the age and design of the plant, the redundancy of auxiliary plant, and luck, the probability of forced outages at time of peak can vary from over 20% down to a few percent.  Nevertheless, power system planners and operators are used to dealing with plant whose output cannot be guaranteed and they therefore include capacity margins in their expansion plans, and provide plant on spinning reserve (synchronised with the system) and on hot and cold standby to cover plant breakdowns.

Once again, the question of reliability becomes a question of degree rather than principle.  If any source of generation has a finite probability of being available for generation at the time of system peak, it has a potential value by reducing the plant capacity that would otherwise have to be constructed.  Obviously, if an ocean energy source has a relatively low probability of being available then the effect that it has on the system peaking capacity requirement will be small.  Analysis must be done carefully and new models will be required:  renewable sources may exhibit a lack of failure diversity, and there may or may not be a correlation between the weather conditions which provide the energy source and the weather conditions which lead to high demands for electricity.

To illustrate the problem, consider a range of wave energy conversions off coast. The weather conditions which lead to high wave energy availability would be brisk winds created by a well-developed depression. In maritime climates, such depressions are often associated with milder weather when heating loads are lower, meaning that wave energy might be highest when demand is relatively low. On the other hand, if the winds also blew on land the cooling effect would increase demand, so that there would be a positive correlation between wave energy output and demand. If the wave converters were widely spread achieving geographic diversity some increased reliability of output would be provided, as it is less likely for the seas to be calm everywhere in the energy recovery reach.

Ocean energy can thus make some contribution to system generating capacity but it has to be carefully modelled and measured against the security of supply criteria laid down by the utilities on the network it is to supply.

Having generalised about large power systems, it is useful to consider how ocean energy devices measure up to the various criteria outlined above.

**Integration of electrical energy from tides**

Existing tidal power developments are all small in relation to the size of the power system to which they are connected [Chapter 3], and their power output can be absorbed without technical difficulty resulting in a fuel saving due to the displacement of thermal plant. These fuel savings would probably include all the overall running costs because the predictability of tidal output would allow steam plant to be programmed to shut down when the state of the tides would allow it. Obviously, as the relative size of a tidal power station increases so will the operational ramifications for thermal plant. A limiting factor, not previously discussed, is a possible difficulty with the high rate of change of power at the ends of the pulse of tidal energy (most acute if tidal output was dropping while consumer demand was increasingly rapidly).

Ebb generation with a single basin tidal power scheme produces the greatest energy, and it should be possible to integrate it with thermal sources more efficiently since it would be possible to shut down some of the latter during the five hour tidal power pulses. The alternative, two-way generation, would either require much shorter runs by the thermal plant or prolonged part load running.

Spring tides can produce typically five to eight times as much energy as leap tides although the power plant may not be designed to capture the full spring tide energy. At any one location the times of high water springs are clustered around particular times [Shaw, 1978], and ideally a site would be chosen where the timing was such that the spring tide peak pulses did not occur during the overnight consumer demand trough. For this reason, the contribution of tidal power to system capacity would have to be carefully measured case by case.

**Integration of electrical energy from ocean thermal sources and waves**

Both ocean thermal and wave generation should be easier to integrate into a power system of predominantly conventional sources than would be the case with tidal generation. Ocean thermal generation should provide a steady base load contribution. If the contribution became large enough to produce a surplus during demand troughs it could be regulated down or stored (for example, by pumped storage) or consumer demand could be stimulated by offering it at an attractively low rate. Efficient integration of wave energy will require accurate predictions of its output at a four to six hour lead time, although this should not be a great problem since wave height changes slowly. The sea is in effect an energy store, as the energy from strong winds is stored in wave patterns which only abate slowly in the ensuing calmer conditions.

## Integration of ocean energy into medium and small electricity systems

In principle the same considerations exist when integrating energy into medium and small electricity systems as into larger systems, although the ocean energy source is likely to be proportionately larger. The quality of supply to the consumer still depends on the size and availability of the various energy sources coupled with the means of demand management, if any. There are a range of techniques available (for example, standby diesel, battery storage, and dump load), the choice depending upon cost and level of sophistication desired.

Fluctuation in the output of an individual device can be significant and the effect on economic outturn substantial. For instance, studies of wind/diesel systems have shown that the diesel compensation for fluctuations in output from the wind generator reduced fuel savings to only about half those that would be expected from the same amount of electricity produced at a steady rate.

We have concentrated on the integration of the output of the energy conversion devices directly with the electricity grid without storage or smoothing. Furthermore, we have said that there are few technical barriers, and that the limitation on its use is principally an economic one, for only when large quantities of ocean energy are envisaged does smoothing and storage become an issue. In such cases, the storage devices should be assessed against the overall system requirement (including the ocean energy devices) for storage, and the assessment will principally once again be an economic one. The storage device will normally have value to the system at large as well as to the integration of the ocean energy device.

## Energy storage

The following storage technologies will be considered:

- pumped storage;

- compressed air;

- gas pipeline for compressed air or hydrogen;

- flywheels;

- batteries; and

- superconducting magnets.

## Pumped storage

Pumped storage is well established as a method of storing energy from an electric power system at times when there is more generating capacity available than is required to meet the consumers' demand. Pumped storage power stations use the principle of a normal hydro-electric power station, where water falls from a high level reservoir, passing through a turbine which drives an electric generator, but with the additional feature that the process can be reversed. This is done by supplying power from the system to the generator so that it becomes a motor driving the turbine, thereby pumping water back up to the top reservoir. The first pumped storage power stations came into use in Europe in the 1920s but the principle was only widely accepted with the development of high capacity reversible pump turbines in the 1960s. Some of the largest stations now exceed 2000MW. The process's turnround efficiency, including hydraulic losses, is generally 75% to 80%.

One constraint on building pumped storage stations has been the need for a suitable site where upper and lower reservoirs can be built with a large vertical but small horizontal separation (in order to reduce friction losses in the shafts and tunnels). Suitable natural sites are rare and in any case may be in remote areas requiring large transmission expenditure. A proposed solution to these difficulties is to locate the lower reservoir underground. Although excavation at substantial depth would add to construction costs, the lower reservoir could be located to provide much higher operating heads than usually occur naturally.

## Compressed air

The technology is available to use compressed air as the storage medium in an underground reservoir, and the principle has been applied at an installation built at Huntorf in Germany during the 1970s. Air is stored in an underground receiver, which can either be a mined cavern or a salt cavity formed by water leaching.

Fig 11-4 on page 13 shows the main components of a compressed air storage system which combines air storage with a gas turbine power generator. However, the energy for air compression, instead of coming from the gas turbine, is taken from the power system at times when relatively cheap energy is available. The power generation uses compressed air from the storage plus fuel to operate the gas turbine and thus drive the generator. For a given power output the gas turbine can be much smaller (approximately one third) than for gas generation alone since it does not have to provide the power for air compression when generating.

## Gas pipeline for compressed air or hydrogen

On large natural gas networks energy is stored by building up the pipeline gas pressure. In the future this technique may be used for combined storage and transmission of ocean energy; for example, a recent study proposal for tidal energy

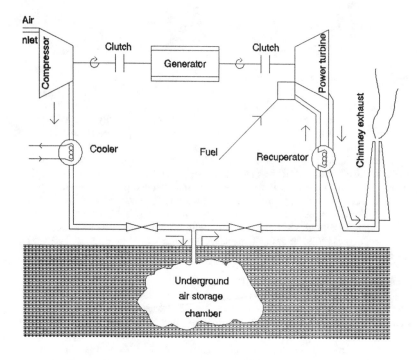

*Fig 11-4 Compressed air storage*

recovery in conjunction with compressed air storage used compressed air as the means of transmission from a compressor driven via a gear-box on the hydraulic turbine shaft via pipes directly to the compressed air storage.

Hydrogen gas is also suitable for combined storage and transmission by pipeline. The hydrogen can be made by electrolysis, and, after transmission to the point of use, can be employed either in an internal combustion engine or directly for heat. The benign nature of the product of combustion (water vapour) makes hydrogen an environmentally neutral fuel.

## Flywheels

Energy storage by flywheel might be applied to small power supply systems to help smooth out short term fluctuations in output conversion devices. For example, in a wind/diesel hybrid generation system the diesel plant will have significantly better fuel consumption if it can avoid starting up during short-term lulls, achieved by connecting a flywheel to the back up diesel in the manner shown in Figure 11-5.

Flywheels can provide energy storage of a few kWh with high in/out efficiency and low losses (ideally the flywheel would be in an evacuated chamber).

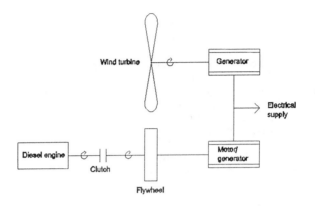

*Fig 11-5 Flywheel storage in wind/diesel system*

## Batteries

Batteries are another means of storage that have been applied to small electrical systems. Battery technology continues to be developed for application to electric transport, but the lead acid battery still predominates. High rates of charge and discharge adversely affect battery life.

## Superconducting magnets

The possibility of electrical superconductivity at achievable temperatures is receiving considerable research effort. However, in addition to the low temperature necessary, large electromagnetic forces would be present, so this method seems to be one for the more distant future.

## Transmission

Before ocean energy can be integrated with a network it requires a transmission connection to get the energy to the point of use. Electricity is one established method of moving energy over long distances, but it should not be assumed that it will always be appropriate. Another method is by pipeline, as has been described briefly in the context of energy storage on pages 13-14. Electrical energy transmission can be either as alternating or direct current.

## Alternating current

Alternating current, generally at either 50Hz (adopted as the European standard) or 60Hz (the North American standard), is the normal method used for medium and large transmission systems.  The main advantages of using AC are:

- voltage levels can be readily changed using transformers;

- interruption of current by switchgear is helped by occurrence of current zeros, even for milliseconds;  and

- construction of machines simplified by the absence of commutators.

If transmission distances are long, difficulties with lack of stability can be experienced if the synchronisation fails between the two ends.  High voltage underground cables are restricted in use to relatively short distances due to their shunt capacitance (which can produce uncontrollable voltage rises) and high cost.

Frequency conversion is possible by using solid state power electronics, for example to couple a variable speed AC machine to a fixed frequency system.

**Direct current**

High voltage direct current is the established method for very long distance overhead transmission and for submarine transmission using cables in order to overcome the drawbacks of AC mentioned above.

DC transmission is also useful when coupling two AC systems if it is necessary to control the amount of power flow between them, as may be the case when output is collected from a number of devices for onward connection.  DC transmission requires complex and expensive terminal equipment, but the cables and lines are simpler and cheaper.  DC technology has not yet been packaged into convenient waterproof robust and compact units for offshore use, but this technology may not be long in coming.

**Demand side management**

Another means of coping with the integration of large amounts of ocean or other renewable energies into electrical power systems would be to develop better demand management systems.  The objective would be to move some customer demand to a time when more renewable energy is available.

The philosophy that the customer's demand must be met in full has arisen partly because, until comparatively recently, it has not been possible to have remote control of such demand according to the priority or importance of individual loads. The technology to provide a selective supply to individual loads according to their importance now exists.   The application of this technology will assist the development of renewable energy sources by helping to make their frequently intermittent nature less of a liability.

The general techniques used in demand side management to modify consumer demand shape in order to assist the operation of generating plant can be divided into three types:  passive;  remote on/off control;  and remote price control.

## Passive

The passive technique uses time of day tariffs in conjunction with time switches in consumer's premises. It is suitable for predominantly thermal power systems where the overall thermal efficiency should rise if more running can be achieved for the efficient, near base load generators. Favourable time of day tariffs (off peak) attempt to achieve this by providing an incentive (cheaper electricity) to consumers if they limit the use of certain appliances (for example, storage and water heaters) to times when the total system demand is low.

The settings on these time switches can only be changed by a supply company employee going to the consumer's site, so that unless only a few large consumers are involved, day to day changes are unpractical and even seasonal changes for domestic consumers may be difficult to justify. As the technology stands the inherent inflexibility of this technique makes it unsuitable for helping with the variations of tidal or wave energy.

## Remote on/off control

This technique differs from the passive in that instead of using time switches to make the cheaper supply available there are switches remotely operated by the supply authority generation control room. The tariff could perhaps guarantee a supply for, say, ten hours per day with at least two hours of supply in any twelve; it is the duty of the control engineer to select times to satisfy the tariff which best suits the available generation. Since the timing of supply to storage loads is not critical the consumer would probably be unaware of changes to the supply availability. The signalling channel can be the mains supply itself or a telephone line or radio (in the UK a hidden, phase-modulated signal is transmitted on the 200 kHz national broadcasting frequency).

Use of the distribution system itself for signalling is not a new idea, having been tried many years since to switch certain specialised loads (for example, street lighting) from a central point. The system (then known as ripple control) was not a great success because effective technology did not then exist for injecting or detecting the signal, but modern electronics has now brought an improvement.

## Remote price control

With remote price control the supply company uses the communication channel to inform the customer of the variation of the price of electricity throughout each day, and the decision to respond to a price change by modifying demand rests with the customer. Although the technology exists the method is not known to have been put into practice, but this is being considered. The price notification could either be predictive or changed in real time. If a two way communication channel was established it could perform meter reading, permit load monitoring and also serve other utilities.

## Conclusion

Ocean energy recovery can be enhanced by newly developed technology used in electrical power generation, storage, transmission and supply.

With large interconnected systems the variable, intermittent or cyclic energy output of some of the ocean sources can be integrated with other forms of generation with comparative ease. Improvements in technology and operational approach may be expected to enhance the value of such sources to both the energy and capacity demand needs of power systems.

Smaller systems, where renewable energy sources form a higher percentage of the generating capacity, have greater difficulty in absorbing it conveniently. Energy storage system and, in future, demand side management, are expected to enhance the value of renewable energy sources.

## References

## Bibliography

T.L. Shaw,    *The role of tidal power stations in future scenarios for electricity storage in UK*. Proc. International Symposium on Wave and Tidal Energy, BHRA Fluid Engineering, Cramfield, Bedford, England. Volume 1, Paper H2 September 1978.


Allen, R.D., and Doherty, T.J., *Dynamic and other secondary benefits of compressed air energy storage*. Proc. International Symposium on the Dynamic Benefits of Energy Storage Plant Operation, Boston, Massachusetts, USA. Session I pp.38-42, May 1984.

CHAPTER 12. FACILITATING TECHNOLOGY FOR FUEl PRODUCTION AND
ENERGY-ENHANCED PRODUCTS

Patrick Takahashi, Member, ASCE
Charles Kinoshita
and Stephen Oney[1]
Joseph Vadus[2]

ABSTRACT

For offshore power production technologies such as ocean thermal energy
conversion (OTEC) to prove competitive in the world energy market, energy storage
media must be utilized to transfer this resource to areas of high energy consumption.
This need has sparked interest in the development of transportable energy carriers and
other products which take advantage of power produced by ocean energy conversion
systems.

Very few studies on facilitating technologies associated with ocean-based fuel
production and energy-enhanced products have been performed in the past because the
emphasis was on production. Ocean energy conversion systems have not yet reached
the commercial stage, this option is on the verge of being utilized. This potential could
be enhanced if ancillary products are synergistically developed. It is this integration
that will make the development of ocean resource technologies economically viable in
the future.

INTRODUCTION

The biggest collection and storage medium for solar energy incident to earth is the
tropical ocean mixed surface layer. The vast potential of this energy reservoir can be
brought into perspective by noting that a quantity of electric power equal to the entire
projected U.S. demand for the year 2000 (about 700,000 MW) could be obtained by
tapping 0.004% of the solar energy contained within the ±10° latitude band flanking
the equator (Dugger, Naef, and Snyder 1981).

While much energy is stored within tropical oceans, the tropics generally are not
noted as centers of industrial development nor as significant energy markets. Thus, for
offshore power production technologies such as OTEC, wave, and tidal power to prove
competitive in the world energy market, it is necessary to formulate means to
redistribute the energy generated by these technologies to areas of high energy
consumption both efficiently and economically. This need has sparked interest in the
development of transportable fuels and other products which take advantage of power
produced by ocean energy conversion systems in their manufacturing process. In fact,
the most recent U.S. Department of Energy review on the concept reported:

> "The greatest potential for OTEC is to supply a significant fraction of the
> world's fuel needs using large plantships to produce hydrogen, ammonia, or
> methanol." (*The Potential of Renewable Energy: An Interlaboratory White
> Paper*, Idaho National Engineering Laboratory, Los Alamos National
> Laboratory,Oak Ridge National Laboratory, Sandia National Laboratory,
> Solar Energy Research Institute, March 1990)

Only a few studies on facilitating technologies associated with ocean-based fuel
production and energy-enhanced products have been performed in the past. This is

[1]Hawaii Natural Energy Institute, University of Hawaii at Manoa, Honolulu, HI 96822
[2]National Oceanic and Atmospheric Administration, Rockville, MD 20852

mainly because, to date, emphasis has been placed on optimizing isolated ocean energy conversion technologies such as OTEC to the point where they can either compete as alternative energy systems in their own right or at least show enough potential to warrant further development. In spite of the substantial effort expended thus far to improve isolated ocean energy conversion systems, they have not reached the commercial stage, more for economic than technical reasons. One major reason is that the isolated technologies have not capitalized on ancillary products or on inputs that might synergistically integrate into these systems. In the opinion of the authors, it is this integration that will make the development of ocean resources economically viable in the future.

In the following sections of this chapter, some of the technologies considered to date, which could form the bases for more integrated systems, are reviewed. The economics associated with these technologies are discussed. A trial case is described to demonstrate the interrelationships necessary for a more integrated and economically viable approach and to identify those areas and technologies that need further development.

## ALTERNATIVE FUELS AND ENERGY-ENHANCED PRODUCTS

The alternative fuels and energy-enhanced products considered in this section fall into two broad categories: (1) resource-based, in which marine biomass is converted into a fuel, a biologically derived product is produced, or a mineral resource is extracted from the sea; and (2) energy-derived, in which the electricity generated from an ocean resource (e.g., thermal gradient, wave, current, tidal power, salinity gradient, marine biomass, or wind) is a major component in the manufacturing of products such as hydrogen, ammonia, or alcohol.

### RESOURCE-BASED OPTIONS

Resource-based options fall into two general groups, biological or mineral. The biological component is itself dependent upon mineral resources, although nutrients within the biogeochemical global ecosystem often are not categorized as mineral resources. The deep waters of all the world's oceans contain the accumulated nutrients of thousands of centuries of growth and decay from terrestrial and surface marine waters. The resurfacing of this nutrient resource in areas of natural upwelling adjacent to coasts accounts for the majority of marine fisheries and bioresources. It is reported that in these naturally upwelled sites, 0.1% of the ocean surface produces 44% of the fish catch (Roels 1980). Maximizing marine productivity requires the judicious use of this resource. More conventional mineral resources within the basins of the world's oceans also represent a vast and largely untapped source of new materials. These materials could be of strategic, international importance because of their potential impact on balancing the unequal distribution of mineral resources which exists in the terrestrial realm.

#### Biological Resources

Farming of aquatic biomass, mainly macroalgae, is one approach of taking advantage of nutrient-rich, pathogen-free, deep ocean waters, and of ocean energy conversion systems. Today, marine biomass is being used for the production of colloidal compounds such as alginic acid, carrageenan, and agar; for human consumption; as an animal feed supplement; and as fertilizers and soil conditioners. The increasing demand for fuels and other high-valued products, coupled with the diminishing ability of land-based resources to meet the growing demand for such products, has stimulated a search for more efficient ways to tap the ocean's biological resources. This section discusses the production of marine biomass and explores its potential for conversion into gaseous or liquid fuels or other high-valued products. Although high-valued, non-energy products from marine biomass may hold greater economical potential in the near-term than fuels, relatively little information exists on the former; therefore, the following discussion focuses primarily on the production of fuels.

Yield studies pertaining to near-shore (along the coast or in tidal flats) and deep-ocean (in rope curtains or in floating lenses) production of macroalgae have been encouraging. Ryther (1979) investigated the growth rates of *Gracilaria sp.* and *Macrocystis pyrifera* and found that the potential yields of these macroalgae species were comparable to or better than most terrestrial crops. North (1987) investigated the potential for open ocean farming of *Macrocystis* and concluded that, while many problems existed relating to large scale farming in the open ocean, technological advancements were capable of overcoming those problems. He also determined that artificial upwelling of deep ocean water was an appropriate fertilizing technique to provide nutrients to enhance algal growth rates. Experiments on land have been reported by Dugan, Cheng, and Takahashi (1987).

A number of thermochemical and biochemical processes could be employed to convert the energy contained in biomass into a more usable form; however, very few technologies are well suited for feedstocks with high (>80%) moisture content, as is typically the case with marine plants. Two conventional technologies are (1) anaerobic digestion, to produce a medium energy content biogas (containing 50%-60% methane with the remainder mostly carbon dioxide) which could be burned directly to generate electrical power, refined into pipeline quality methane, or converted into methanol using existing technology; and (2) fermentation, to produce alcohols or other high-valued organic compounds such as acetone. An exciting new development reported by Manarungson, Mok, and Antal (1990) investigates the gasification of high-moisture content marine algae to produce hydrogen using supercritical water.

In comparison with terrestrial biomass, little is known of the microbiology and biochemistry of anaerobic digestion of marine macroalgae. However, the limited research that has been performed on bioconversion of marine biomass into methane has produced encouraging results—rapid and nearly complete conversion, and process stability, seem to be inherent in biochemical conversion of marine biomass. For example, Chynoweth, Fannin, and Srivastava (1987) studied the potential for biological gasification of marine algae (primarily *Macrocystis*) and concluded that marine algae were excellent feedstocks for the production of methane. They found that anaerobic digestion of *Macrocystis pyrifera* yielded more than 80% of the theoretically attainable methane, the highest yield ever reported for any non-waste biomass. Of the many marine plants available for bioconversion into methane, two subgroups of the brown kelp (Phaeophyta) phyla, *Macrocystis pyrifera* (best in temperate seas or for cultivation in cool, upwelled water) and *Sargassum* (best in tropical seas), hold the most promise. A schematic for a hypothetical marine biomass-to-methane bioconversion system is shown in Fig. 1. In this fuel production scenario, 335 tons (dry basis) of marine biomass per day would produce 19,900 tons (one million GJ) of methane annually.

A number of potential by-products could complement the production of methane from macroalgae. For example, more than twenty potential products which could be co-produced with methane from kelp have been identified in the literature (Bird 1987). Bird (1987) performed a cost analysis on energy from marine biomass which suggested that its economic feasibility lies in limiting energy costs by exploiting potentially available ocean energy systems and in enhancing methane production with revenues from by-products. His calculations indicated that co-production of energy and chemical products could more than double gross sales receipts over producing only methane.

Conversion of biomass into methane is not as energy intensive as the electricity-based options discussed in the next section. Therefore, in most of the biomass-to-methane systems proposed to date, the bioconversion plant was assumed to be located on land and not integrated with an ocean energy electrical generating facility.

However, marine biomass production could be greatly enhanced by using the nutrient-rich cold water upwelled by an OTEC facility, which suggests an advantage in integrating biomass production with OTEC power generation. In addition, the lower "land" cost and relative ease of harvesting and transporting the feedstock to a nearby floating platform would provide economic advantages. Thus, while the Longwell study (1989) showed that methanol from terrestrial biomass was not economically attractive, methanol from marine biomass could have a more favorable future.

As an alternative to the production of methane, marine biomass could be converted into alcohol fuels via fermentation. Very little research has been performed in this area, and cost projections (to be discussed in a later section) are quite preliminary. The process of converting marine biomass into ethanol would resemble that employed for grains. Conversion of marine biomass would probably require more energy input due to the higher moisture content in the feedstock and may require an initial hydrolysis step to break down the complex hydrocolloids into fermentable oligosaccharides (Bird 1987). Overall conversion would be strongly influenced by the proportion of soluble carbohydrates to total carbohydrates, which, in marine biomass, can vary from 20% to 70%.

Other biofuel farming concepts have surfaced in the past, as for example, the report by Wagener (1981) of the European Community to utilize coastal deserts. Since no fertile land nor fresh water need is anticipated, such a system could be attractive in niche areas.

Mineral Resources

Offshore oil and natural gas represent developed energy resources, but methane hydrate deposits and deep oil and gas reserves are two examples of fossil-fuel sources of oceanic origin that are worthy of mention. Presently, oil drilling technology in the open oceans is limited to ~1000 to 2000 m due to problems with riser-pipe performance beyond those depths (Cruickshank 1990). Methane hydrate, a clatherate of water molecules surrounding methane gas, exists in significant concentrations throughout the ocean floor. It has not been tapped as a source of fuel thus far because of the lack of technology associated with the mining of the crystals and their conversion into usable methane.

The technology associated with offshore processing of ferromanganese nodules and crusts recovered from the ocean floor is still in a developmental stage, and much more research is required before the processing of these ores is ready for commercialization (Cruickshank 1990). Nevertheless, the potential benefits that can be accrued by combining deep ocean mining with ocean energy recovery have sparked interest in such an integrated approach. The best sites for many ocean energy systems such as OTEC are in proximity to sites for deep ocean mining (Grote 1981). Processing in the region of mining could greatly reduce transportation and waste disposal costs associated with the pyrometallurgical process used in refining manganese nodules into cobalt, copper, and nickel. The water needed for cooling in ore processing could be taken directly from the adjacent sea water, or in a more integrated approach, from the cold water pumping stream of an OTEC plant. Utilizing the cold deep ocean waters in this way could accomplish two goals: (1) cooling in the metallurgical process and (2) heating of nutrient-rich waters for subsequent use in open ocean mariculture, which is important because, otherwise, the cold deep ocean water might sink below the euphotic zone before the nutrients could be consumed by marine biomass.

The ocean is a storehouse of minerals, and sodium chloride continues to be extracted through an evaporative process. Bromine and magnesium until recently were produced from sea water on an industrial scale. The Japanese have conducted

considerable research on extracting uranium from sea water. However, the projected payback was not sufficient for commercial applications.

The large volume of OTEC fluids has sparked recent interest on the recovery of gold and other trace metals. The value of gold passing through a one megawatt OTEC power plant exceeds $50 million each year. Concepts such as synthesis of special molecules, as for example, (24-pyrimidium Crown 6), have been funded by the Pacific International Center for High Technology Research, and genetic engineering micro-organisms to adsorb specific elements have been suggested.

ENERGY-DERIVED OPTIONS

Feedstock conversion processes that require substantial input of power could capitalize on energy from the ocean if the cost of the energy were relatively low. Examples of feed materials that potentially might be used in power-intensive processes to produce transportable fuels and energy-enhanced products include sea water, marine biomass, seabed ores, coal, and municipal wastes. Sea water could be converted into hydrogen and oxygen via ocean-energy powered electrolysis; these, in turn, could be used in the processing of other products such as ammonia. Marine biomass could be converted into liquid fuels such as methanol and other energy-enhanced products (e.g., fertilizers) and chemicals. Seabed ores mined near the power generating facility could be processed on site, utilizing the energy produced for the smelting and refining processes in converting the raw ores into commercial grade metals. Such feed materials have the advantage of potentially being extracted in proximity to, or produced in conjunction with, an ocean-based power system; therefore, high shipping costs would not be incurred in transporting the material from distant locations to the plant site. Coal and municipal wastes and other terrestrial feedstocks have also been considered for use with ocean energy conversion systems in the production of methane or methanol. Such feedstocks would have to be transported from land to power production facilities in the ocean where they would then be converted into other products. Thus, it would be advantageous to integrate marine biomass plantations with ocean platform facilities.

Process diagrams for three of the fuel production options mentioned above, hydrogen, ammonia, and methanol, as proposed by Avery, Richards, and Dugger (1985), are shown in Fig. 2, 3, and 4. (The sizes of the systems shown in Fig. 2, 3, and 4 relate to a common production unit, one million GJ, net heating value, of fuel produced.) Each of these fuel options employs a common base process—electrolysis of desalinated water, using electricity generated by an ocean energy conversion plantship, to produce gaseous hydrogen and oxygen. Additional steps, as described below, are added to the base process to produce the desired product. (Note, in Fig. 2, 3, and 4, $MW_e$ and $MW_t$, denote electrical and thermal energy transfers, respectively.)

Hydrogen

Hydrogen is regarded by many to be the best long-term alternative to fossil fuels—it is renewable when produced from renewable resources, and environmentally benign. Avery, Richards, and Dugger (1985) and Avery (1988) investigated the production of hydrogen via water electrolysis in 50-400 MW OTEC plantships. Their analyses indicated that power could be generated on plantships of that size range at costs low enough to permit fuels and other products (ammonia, fertilizers, etc.) to be manufactured on board and delivered to land-based users at prices approaching those for products from conventional methods of production that utilize fossil fuel or nuclear power.

In the OTEC-hydrogen plantship scenario (Fig. 2), 64 MW (net power production) from an ocean energy conversion facility could produce 8270 tons (one million GJ, net heating value) of hydrogen per year. The low density and low boiling point of the hydrogen produced would demand special handling and storage measures at additional cost. Hydrogen, once liquefied, would be shipped to market in tankers designed for low density cryogenic liquids or embedded into metal hydrides or another convenient carrier, and then regenerated as needed.

Ammonia

Ammonia exists in liquid form at moderate pressures and temperatures, is much easier to handle than hydrogen, and contains more hydrogen per unit volume than metal hydrides or than liquid hydrogen itself. In addition, ammonia is a good fuel in its own right and is a base for the production of fertilizers and industrial chemicals. Avery (1985, 1988) argued that, for subsequent transportation and energy utilization, it would be more economical to convert the hydrogen produced in a plantship into ammonia. Richards and Henderson (1981) performed a survey of previous work on the use of ammonia as a fuel and noted that much of the distribution infrastructure and markets for ammonia already existed. They concluded that ammonia produced by ocean energy plantships could become an economical fuel in the near future.

In the ammonia production scenario envisioned by Avery, Richards, and Dugger (1985), electricity from an ocean energy conversion facility would be used to produce nitrogen via air liquefaction. The nitrogen would then be combined with the hydrogen generated as described earlier and catalytically converted into ammonia. Fig. 3 shows a 51 MW (net) electrical generating facility capable of producing 53,800 tons (one million GJ) of ammonia annually.

Methanol

Methanol has been acclaimed as the best near-term alternative to conventional ground transportation fuels, gasoline and diesel, from the standpoints of cost, performance, and ease of handling and distribution. The potentially large demand for methanol and its likelihood of being accepted into the market, make methanol a strong candidate for production in ocean-based manufacturing plants. In the ocean-based methanol production scenario, a carbonaceous feedstock (coal, biomass, biogas, or the like) would be partially oxidized with oxygen produced by electrolysis as described above into carbon monoxide and other gases. The carbon monoxide would then be combined with hydrogen from the electrolysis plant and catalytically converted into methanol. In the coal-to-methanol scenario (Fig. 4), 14 MW (net) of electricity from the plantship combined with 107 tons of coal per day (1.24 kg/s) would produce 47,400 tons (one million GJ) of methanol annually, thus yielding 1.3 tons of methanol per ton of coal. By comparison, land-based coal-to-methanol plants (which would obtain most of the required hydrogen by reacting coal with water and oxygen and which would employ air liquefaction to produce the necessary oxygen) would probably be capable of yielding only 0.5 to 0.7 tons of methanol per ton of coal (Avery and Richards 1982; Avery, Richards, and Dugger 1985).

Marine biomass could be converted into methanol in much the same manner as coal. Takahashi et al. (1990) reported that the yield of methanol from gasified biomass could be increased by more than 100% by adding enough hydrogen to the intermediate gas stream to obtain the stoichiometric hydrogen:carbon monoxide ratio for methanol synthesis; comparable methanol yield increases could be achieved with marine biomass by adding hydrogen produced by an ocean-based electrolysis plant to the product gas.

If desired, the methanol produced by the plantship could be converted into gasoline (e.g., via the Mobil process). Alternatively, the methanol production step could be bypassed through the use of special catalysts (Parmon 1989).

Non-Fuel Products

Power from ocean resources can also be used synergistically in the manufacturing of various non-fuel products. For example, several have suggested that ocean-derived electricity be used for alumina-to-aluminum reduction. Combining the reduction process with ocean energy conversion is attractive because (1) the reduction process, which is energy intensive (approximately 15.4 kWh/kg aluminum is required for the electrolysis cells and ancillary demand), could fully take advantage of ocean-derived electricity; and (2) ocean-based alumina reduction plants could facilitate reduction and minimize shipping costs by being located along alumina shipping routes, e.g., near Hawaii, Puerto Rico, tropical islands, South American coasts, or Australia (Dugger, Naef, and Snyder 1981)

PRELIMINARY ECONOMICS

Commercially speaking, ocean energy conversion, farming of marine biomass, and the extraction of chemicals and minerals from the ocean are either nonexistent or in their infancy. Most of the economic evaluations of ocean resource recovery systems performed to date have been confined to alternatives which produce only electricity. Therefore, there is little information upon which the economic feasibility of producing fuels and energy-enhanced products in ocean-based systems, especially integrated systems, can be accurately assessed.

Estimated production costs for some of the single-fuel options discussed earlier are tabulated in Table 1. (The figures in Table 1 represent the estimated production costs at the time of the respective studies; no attempt is made here to adjust those figures to a common base period.)

The anticipated cost of producing pipeline quality methane from kelp ($8/GJ) is substantially higher than the current price of natural gas in the U.S., and the cost of producing ethanol from kelp ($20/GJ), while comparable to that from woody biomass, is significantly higher than the current cost of converting starch- or sugar-based feedstocks into ethanol. The anticipated cost of the feedstock delivered to the processing plant, ~$2/GJ to $6/GJ (Bird 1987), represents a large portion of the projected overall cost of producing methane from marine biomass. Therefore, substantial reductions in the cost of cultivating, harvesting, and delivering the biomass to the most optimistic levels cited in the literature could help to make the cost of producing methane from marine biomass competitive with current terrestrial methods.

The capital costs of the ocean thermal energy conversion portions of the hydrogen, ammonia, and methanol plantships used in calculating the figures presented in Table 1 were approximate and widely uncertain (Avery, Richards, and Dugger 1985). Since the OTEC portions represent a major component in the overall cost of producing all three products, the projected costs tabulated in Table 1 should be considered as preliminary. Even though the assumed cost for electricity was very low, the projected cost of producing liquid hydrogen in an OTEC hydrogen plantship, $24/GJ, is much higher than that obtainable from land-based plants employing steam reforming of methane. Nevertheless, many are optimistic about the long-term feasibility of producing such fuels. Avery and Richards (1982) and Avery (1985) believed that OTEC-derived fuels delivered to world ports could become cost competitive with both motor vehicle fuels derived from petroleum and synthetic fuels produced in land-based plants. O'Sullivan and Camara (1981) concurred, suggesting that cost-competitive electric power could be regenerated from OTEC produced ammonia and methanol if those fuels were used in conjunction with molten carbonate fuel cells.

The Pacific International Center for High Technology Research is working with the Hawaii Natural Energy Institute on a contract to determine the feasibility of producing

OTEC hydrogen. Related experiments will be performed at the Natural Energy Laboratory of Hawaii.

Since hydrogen is a key element in the production of ammonia and methanol, significant reductions in the cost of producing hydrogen could, in turn, make the production of ammonia and methanol by ocean energy conversion plants competitive with current, land-based processes which use natural gas (Avery 1985, 1988). Indeed, recent advances in electrolysis for large scale production of hydrogen (Kincaid and Murray 1981; Nuttall 1981; Crawford and Benzimra 1986; Lodhi 1987) could significantly enhance the economic feasibility of all three ocean-based products.

## SYSTEM INTEGRATION—A TRIAL CASE

Key to commercialization of ocean energy resource recovery is the integrated use of cold deep ocean water to produce energy through non-polluting conversion technologies and to provide nutrients for mariculture in such a manner that synergistic cost savings result. An example of such an integrated approach, based on an OTEC cycle, is shown in Fig. 5 (Vadus 1981). Here, cold upwelled water is used to generate electricity, to enhance marine biomass production and aquaculture, and to provide air conditioning and refrigeration within the integrated facility. The electricity produced by the OTEC plant is used to process and refine various fuels and other chemical and metal products, to support related maricultural and aquacultural activities, and, if practicable, for export. Fresh water, produced from sea water via distillation or by adding a bottom cycle to the OTEC plant, is used for domestic purposes and for the production of hydrogen or ammonia, as described earlier. Marine biomass farmed at the facility would be converted into methane, methanol, fertilizers, or animal feed, and aquaculture would provide a variety of food products.

Such an integrated approach is more likely to succeed commercially than the isolated, single-product alternatives discussed earlier, but the integrated approach is also much more complicated. Therefore, the concept needs to be developed in stages. Work is currently underway to design, build, and test 160-kW closed-cycle, 165-kW open-cycle, and 500-kW hybrid-cycle OTEC prototypes at the Natural Energy Laboratory of Hawaii by the mid-1990s.

Once the concept has been demonstrated at the smaller scale, the technology could be upscaled to the precommercial level (~5 to 10 MW), beginning with the cold and warm water conveyance subsystem, followed by the power generation subsystem, and then the mariculture component. Techniques to control the movement of upwelled water within the euphotic zone long enough for nutrient uptake by macroalgae could be researched and demonstrated at that stage. The successful co-production of electricity and marine biomass in such an integrated fashion would provide (1) information vital to the design and operation of a full-scale (~50 MW) power generation/mariculture facility, and (2) confidence in the technical and economic viability of the concept to help secure financing for a commercial venture. Since the production of many of the fuel and energy-enhanced products mentioned earlier relies heavily on electricity generated from ocean resources or on marine biomass, a successful demonstration of electricity and biomass production at the precommercial scale would also enable many of those product options to be considered for incorporation into a commercial system.

## REQUIRED TECHNOLOGY

Existing technologies and those which must be developed to facilitate the production of some of the previously mentioned products are summarized in Table 2. Many of the manufacturing processes listed are well established on land. In most cases, those technologies could be easily adapted to ocean-based systems. Likewise, many of the isolated ocean energy conversion technologies discussed earlier (OTEC and others) can be considered as virtually developed from the technological standpoint (although, in most cases, the economics associated with the individual systems appear poor). The main challenges which remain are the coupling of the two technologies—energy conversion and the manufacturing of an energy-enhanced product—and the integration of complementary processes into economically viable enterprises.

To date, efforts to cultivate kelp in the open ocean have been largely unsuccessful due to various engineering problems associated with large scale marine farming (Neushul 1987). Solutions to these and biological problems must be found before kelp can be successfully grown in the open ocean on a commercial scale. Even though relatively little research has been performed on optimizing the yield of macroalgae, the photosynthetic productivity of many strains of macroalgae are much higher than those of most terrestrial biomass. Indeed, additional basic research to understand the genetics and the physiological requirements of macroalgae and applied research (e.g., establishing hybridization and breeding programs) could result in yield increases comparable to the large increases achieved by major land-based crops in the past few decades.

There is general agreement that upwelled deep ocean water represents a vast potential source of nutrients for marine biomass; however, the increase in yield that might accrue by using this resource is uncertain. Madenjian and McKinley (1989, 1990a, 1990b) have addressed that issue by modeling the incorporation of deep ocean nutrients into marine biomass. Additional research is needed to quantify the benefits associated with the integration of deep ocean upwelling and marine biomass production so that sound decisions can be made to commercialize the concept. Emphasis must also be placed on improving marine biomass harvesting (North 1987), which would likely be the most costly step in providing biomass feedstock to a bioconversion facility.

Additional research on evaluating the effect of feed composition and inocula development to optimize anaerobic digestion of kelp to produce methane (Chynoweth, Fannin, and Srivastava 1987) and scaling-up the technology to formulate and evaluate appropriate materials handling, process control, and effluent utilization procedures, are needed. A major technical bottleneck that often confronts thermochemical conversion processes is the high moisture content typically in the harvested biomass. In the conversion of lignocellulosic feedstocks into methanol, the energy input required to transform biomass into a synthesis gas increases dramatically as the amount of moisture in the reactants increases (Wang, Kinoshita, and Takahashi 1990). The same would be true for marine biomass. Additional research on novel processes which can utilize wet feedstocks, such as pyrolytic reaction of macroalgae in supercritical water (Manarungson, Mok, and Antal 1990), is needed to improve the feasibility of producing fuels from marine biomass.

So far, the production of hydrogen via electrolysis of water with potassium hydroxide solutions has been performed only on a relatively small scale. The high efficiencies (85%-90%) assumed in the literature (e.g., Avery, Richards, and Dugger 1985) must be demonstrated on a commercial scale before the technology can be considered viable and before the economics of such processes can be assessed accurately. Interfacing the hydrogen production facility with the ocean electrical generating plant, storing the liquid hydrogen, and transporting it to market in an economical manner are also major challenges in this option.

The technology to produce ammonia via catalytic reaction of coal or natural gas already exists. The problems associated with the production of ammonia on ocean facilities relate primarily to feedstock transportation and storage. Synthesizing ammonia from hydrogen obtained via electrolysis and nitrogen obtained via air liquefaction poses problems similar to those noted for hydrogen production. Unlike many of the other products being discussed, the infrastructure for transporting, storing, and distributing ammonia already exists and could easily be adapted to ocean-based energy systems (Avery, Richards, and Dugger 1985).

Much of the infrastructure for transporting coal is already in place; hence, while the high cost of transporting and storing coal poses a significant obstacle in the economical production of methanol from coal, there are very few technical problems associated with that option. Many believe that methanol produced from terrestrial biomass is the most feasible near-term alternative to conventional, fossil-based, transportation fuels. Efforts are underway to scale-up biomass gasification to optimize biomass-to-methanol production (Takahashi et al. 1990). Much of the technology being developed for terrestrial biomass could be readily transferred to marine systems. However, certain properties (e.g., moisture and ash contents) of some marine biomass species are substantially different from those of woody biomass; therefore, some research on thermochemical conversion of marine biomass into methanol is needed. The alternative process of biochemically converting marine biomass into methane, then steam reforming or partially oxidizing (with oxygen produced by electrolysis of water) the methane into a synthesis gas for conversion into methanol should also be examined and compared with the solely thermochemical approach.

Refining of alumina via the Hall-Heroult process is well developed. The major concern in performing this operation on the ocean is a potential loss in production efficiency associated with misalignment of the molten aluminum cathode and carbon anode due to wave-induced plant motion (Dugger, Naef, and Snyder 1987); however, recent technological breakthroughs could alleviate this concern.

With regard to the refining of cobalt-rich crusts, substantial research and technological development are still needed in mining and refining. The national ocean basin Marine Mineral Technology Center, funded by the U.S. Department of the Interior at the University of Hawaii, is developing these technologies.

The ultimate enhancement could well be the process of induced upwelling, as reported by North (1989), Takahashi and Trenka (1989), and Takahashi and Yuen (1989), whereby mega-pods of OTEC facilities bring deep-water fluids to the surface, primarily to absorb carbon dioxide from the atmosphere to remediate the greenhouse effect. As carbon dioxide itself is brought to the surface, the material balance will depend on biological growth and the possible requirement of minerals to aid in sequestration. This concept is a debatable one at this time and will require considerable investigation.

CONCLUSIONS

The oceans as a source of fuel and energy-enhanced products are attractive for several reasons. First, there are resources in the oceans that cannot be produced anywhere else in the desired quantities, nor can acceptable substitutes be found. Second, ocean resources can be extracted at a lower cost and more conveniently than their terrestrial counterparts. Third, tapping ocean resources would probably have the least environmental impact.

For these reasons, ocean thermal energy conversion offers one of the best opportunities to use the oceans. And if the OTEC concept is ever to develop into a major energy form, fuels, minerals and chemicals will be the primary reasons why. Many tens of thousands of ocean energy megawatt equivalents can someday be powering floating cities and industrial platforms for strategic metals, transportation fuels, chemicals and other products. The commercial success of these enterprises will almost surely depend on the integration of system packages.

For example, a grazing platform to harvest and process seabed ores, using OTEC electricity, and returning wastes to the ocean floor, could well turn out to be the only acceptable option for strategic metal production. Likewise, a marine biomass plantation encircling a floating methanol production facility in the open ocean utilizing upwelled deep ocean fluids creating new fisheries while possibly enhancing the atmosphere, would have attractive prospects.

During the next decade much of the science and engineering research can be simulated at land-based operations such as at the Natural Energy Laboratory of Hawaii and the Shikoku laboratory of the Japan Marine Science and Technology Center, both of which feature pipes which bring deep, cold water to the surface. However, prototype experiments will also need to be performed on the open ocean to test concepts and gain the confidence of the financial sector.

The National Science Foundation sponsored a workshop in 1989 on "The Ocean Enterprise," where recommended was a series of OTEC related projects leading to a 500-MW ocean mineral platform combining OTEC and seabed ore processing by the year 2000 and a 1000-MW Pan American complex featuring multinational cooperation and a full range of co-products at a cost of $10 billion by 2010 (Ross and Daily 1989). If global warming becomes a serious matter, nuclear power plants continue to face societal acceptance problems, anything close to a minerals crisis occurs, and the commercial equivalent of the National Aerospace Plane--which will be powered by liquid hydrogen--flies, then by year 2020 OTEC should be in a prime position to provide fuels and minerals from the sea.

## ACKNOWLEDGEMENTS

The authors gratefully acknowledge the valuable contributions to this investigation by the following individuals: Michael Cruickshank, Marine Minerals Technology Center, University of Hawaii; Hans Krock, JKK Look Laboratory, University of Hawaii; Kelton McKinley, Hawaii Natural Energy Institute, University of Hawaii; and the staff of the Pacific International Center for High Technology Research. Finally, this chapter is written in tribute to William H. Avery of the Johns Hopkins University, whom we consider to be the "Father of Fuels from the Sea." Dr. Avery provided considerable data for which the authors are indebted. Thank you, Bill!

## REFERENCES

Avery, W.H.   1980.   Grazing OTEC plantships technical status potential products and costs.  In *The National Conference on Renewable Energy Technologies Proceedings*. December 7-11.

Avery, W.H., and D. Richards.  1982.  OTEC methanol.  Presented at The Oceans '82, MTS-IEEE Conference and Exposition, September 20-22.

Avery, W.H.  1985.  Fuels from the oceans via ocean thermal energy conversion (OTEC). Presented at The Ocean Space Utilization Conference, Tokyo.

Avery, W.H., D. Richards, and G.L. Dugger.  1985.  Hydrogen generation by OTEC electrolysis, and economical energy transfer to world markets via ammonia and methanol.  *Int. J. Hydrogen Energy*, 10(11):727-736.

Avery, W.H.  1988.  A role for ammonia in the hydrogen economy.  *Int. J. Hydrogen Energy*, 13(12):761-773.

Bird, K.T.  1987.  Cost analyses of energy from marine biomass.  *Seaweed Cultivation for Renewable Resources*.  Elsevier, New York,  pp. 327-350.

Bird, K.T., and P. Bensen.  1987.  Conclusions and recommendations.  *Seaweed Cultivation for Renewable Resources*.  Elsevier, New York,  pp. 369-371.

Chynoweth, D.P., K.F. Fannin, and V.J. Srivastava.  1987.  Biological gasification of marine algae.  *Seaweed Cultivation for Renewable Resources*.  Elsevier, New York, pp. 285-303.

Crawford, G.A., and S. Benzimra.  1986.  Advances in water electrolyzers and their potential use in ammonia production and other applications.  *Int. J. Hydrogen Energy*, 11(11):691-701.

Cruickshank, M.  Private communication with S.K. Oney, 1990.

Dugan, G.L., D.H. Cheng, and P.K. Takahashi. 1987. Induced mixing characteristics for algal biomass enhancement. In *Energy from Biomass and Waste XI Conference Proceedings*. Institute of Gas Technology. Florida.

Dugger, G.L., F.E. Naef, and J.E. Snyder. 1981. Ocean thermal energy conversion. *Solar Energy Handbook*. McGraw-Hill, New York, pp. 19-1-53.

Grote, P.B. 1981. OTEC power application to deep ocean mining. In *The 8th Ocean Energy Conference Proceedings*. The Marine Technology Society, Washington D.C., pp. 675-678.

Idaho National Engineering Laboratory, Los Alamos National Laboratory, Oak Ridge National Laboratory, Sandia National Laboratory, Solar Energy Research Institute. 1990. The potential of renewable energy, an interlaboratory white paper. Prepared for the Office of Policy, Planning and Analysis, U.S. Department of Energy. Published by the Solar Energy Research Institute.

Kincaid, W.C., and J.N. Murray. 1981. Advanced alkaline electrolysis systems for OTEC. In *The 8th Ocean Energy Conference Proceedings*. The Marine Technology Society, Washington D.C., pp. 685-693.

Lodhi, M.A.K. 1987. Hydrogen production from renewable sources of energy. *Int. J. Hydrogen Energy*, 12(7): 461-468.

Longwell, J.P. 1989. U.S. production of liquid transportation fuels. National Academy Press, Washington, D.C.

Madenjian, C., and K.R. McKinley. 1989. Nitrogen biogeochemistry within artificial cold-core cyclonic structures: a preliminary time-dependent biological model. In *The First International Workshop on Artificial Upwelling and Mixing in Coastal Waters Proceedings*. Keelung, Taiwan, ROC.

Madenjian, C., and K.R. McKinley. 1990. Nitrogen dynamics within oceanic cold-core rings: evidence for the consumer interference effect. *Limnol. Oceanogr.* (submitted).

Madenjian, C., and K.R. McKinley. 1990. Nitrogen dynamics within artificial and natural oceanic cold-core ring structures. Part I: a time-dependent biological model. *J. Mar. Res.* (submitted).

Manarungson, S., W.S. Mok, and M.J. Antal. 1990. Hydrogen production by gasification of glucose and wet biomass in supercritical water. In *World Hydrogen Energy Conference #8 Proceedings*. Honolulu and Waikoloa, Hawaii.

Mencher, F.M., and R.B. Spencer. 1981. OTEC energy transference via biomass: the pros and cons. In *The 8th Ocean Energy Conference Proceedings*. The Marine Technology Society, Washington D.C., pp. 695-698.

Murata, D. and M. Keller. 1977. *Biomass energy for Hawaii Vol. IV*.

Neushul, P. 1987. Energy from marine biomass: the historical record. *Seaweed Cultivation for Renewable Resources*. Elsevier, New York.

North, W.J. 1987. Oceanic farming of *Macrocystis*, the problems and non-problems. *Seaweed Cultivation for Renewable Resources*. Elsevier, New York, pp. 39-67.

North, W.J. 1989. Biological opportunities associated with large offshore platforms. In *Oceans '89 Proceedings*. Seattle.

Nuttall, L.J. 1981. Advanced water electrolysis technology for efficient utilization of ocean thermal energy. In *The 8th Ocean Energy Conference Proceedings*. The Marine Technology Society, Washington D.C., pp. 679-683.

O'Sullivan, J.B., and E.H. Camara. 1981. Application of molten carbonate fuel cells to OTEC-generated fuels. In *The 8th Ocean Energy Conference Proceedings*. The Marine Technology Society, Washington D.C., pp. 699-704.

Parmon, V. 1989. *Hydrogen Photo-Production Workshop III Summary*. Hawaii Natural Energy Institute, Hawaii.

Richards, D., and R.W. Henderson. 1981. OTEC produced ammonia—an alternative fuel. In *The 8th Ocean Energy Conference Proceedings*. The Marine Technology Society, Washington D.C., pp. 665-673.

Roels, O.A. 1980. From the deepsea: food, energy and fresh water. Mechanical Engineering.

Ross, D.A., and J.E. Dailey. 1989. *The Ocean Enterprise Workshop*. National Science Foundation, Washington, D.C., pp. 19.

Ryther, J.H. 1979-80. Fuels from marine biomass. Oceanus 22(4):48-58.

Takahashi, P.K. and A. Trenka, 1989. Status of ocean thermal energy conversion technology in 1989. *ISES Solar World Congress*. Kobe, Japan.

Takahashi, P.K. and P.C. Yuen. 1989. Ocean resource development in Hawaii. In *Oceans' 89 Proceedings*. Seattle.

Takahashi, P.K., D.R. Neill, V.D. Phillips, and C.M. Kinoshita. 1990. Hawaii: an international model for methanol from biomass. *Energy Sources* 12(4):421-428.

Vadus, J.R. 1981. Ocean technology in the 21st century. U.S.-Japan Symposium on Ocean Technology in the 21st Century. Nihon University. Tokyo, Japan.

Vadus, J.R. 1990. Hawaii: poised for the 21st century. *Proceedings of WHEC TREK*. World Hydrogen Energy Conference #8, Honolulu, Hawaii.

Wagener, K. 1981. Energy from marine biomass: methane production by mariculture on land. *Final Report of the Commission of the European Communities*.

Wang, Y., C.M. Kinoshita, and P.K. Takahashi. 1990. Chemical equilibrium computations for gasification of biomass to produce methanol. In *Energy from Biomass and Wastes XIV Conference Proceedings*. Institute of Gas Technology. Florida.

Wise, C.E. 1979. 'Free' and renewable energy sources. *Mach. Des.* 49(12):22-28.

# SUBJECT INDEX
## Page number refers to first page of paper.

# AUTHOR INDEX
Page number refers to first page of paper.

309